T0291813

CAMBRIDGE LIBRARY COLLECTION

Books of enduring scholarly value

Mathematical Sciences

From its pre-historic roots in simple counting to the algorithms powering modern desktop computers, from the genius of Archimedes to the genius of Einstein, advances in mathematical understanding and numerical techniques have been directly responsible for creating the modern world as we know it. This series will provide a library of the most influential publications and writers on mathematics in its broadest sense. As such, it will show not only the deep roots from which modern science and technology have grown, but also the astonishing breadth of application of mathematical techniques in the humanities and social sciences, and in everyday life.

Principles of Geometry

Henry Frederick Baker (1866–1956) was a renowned British mathematician specialising in algebraic geometry. He was elected a Fellow of the Royal Society in 1898 and appointed the Lowndean Professor of Astronomy and Geometry in the University of Cambridge in 1914. First published between 1922 and 1925, the six-volume *Principles of Geometry* was a synthesis of Baker's lecture series on geometry and was the first British work on geometry to use axiomatic methods without the use of co-ordinates. The first four volumes describe the projective geometry of space of between two and five dimensions, with the last two volumes reflecting Baker's later research interests in the birational theory of surfaces. The work as a whole provides a detailed insight into the geometry which was developing at the time of publication. This, the first volume, describes the foundations of projective geometry.

Cambridge University Press has long been a pioneer in the reissuing of out-of-print titles from its own backlist, producing digital reprints of books that are still sought after by scholars and students but could not be reprinted economically using traditional technology. The Cambridge Library Collection extends this activity to a wider range of books which are still of importance to researchers and professionals, either for the source material they contain, or as landmarks in the history of their academic discipline.

Drawing from the world-renowned collections in the Cambridge University Library, and guided by the advice of experts in each subject area, Cambridge University Press is using state-of-the-art scanning machines in its own Printing House to capture the content of each book selected for inclusion. The files are processed to give a consistently clear, crisp image, and the books finished to the high quality standard for which the Press is recognised around the world. The latest print-on-demand technology ensures that the books will remain available indefinitely, and that orders for single or multiple copies can quickly be supplied.

The Cambridge Library Collection will bring back to life books of enduring scholarly value (including out-of-copyright works originally issued by other publishers) across a wide range of disciplines in the humanities and social sciences and in science and technology.

Principles of Geometry

VOLUME 1:
FOUNDATIONS

H.F. BAKER

CAMBRIDGE
UNIVERSITY PRESS

CAMBRIDGE UNIVERSITY PRESS

Cambridge, New York, Melbourne, Madrid, Cape Town, Singapore,
São Paolo, Delhi, Dubai, Tokyo, Mexico City

Published in the United States of America by Cambridge University Press, New York

www.cambridge.org
Information on this title: www.cambridge.org/9781108017770

© in this compilation Cambridge University Press 2010

This edition first published 1922
This digitally printed version 2010

ISBN 978-1-108-01777-0 Paperback

PRINCIPLES OF GEOMETRY

CAMBRIDGE UNIVERSITY PRESS

C. F. CLAY, Manager

LONDON : FETTER LANE, E.C. 4

NEW YORK : THE MACMILLAN CO.
BOMBAY
CALCUTTA } MACMILLAN AND CO., Ltd.
MADRAS
TORONTO : THE MACMILLAN CO. OF
CANADA, Ltd.
TOKYO : MARUZEN-KABUSHIKI-KAISHA

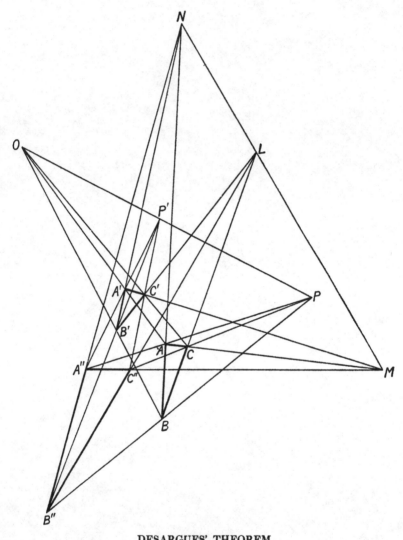

DESARGUES' THEOREM

(see p. 6)

PRINCIPLES OF GEOMETRY

BY

H. F. BAKER, Sc.D., F.R.S.,

LOWNDEAN PROFESSOR OF ASTRONOMY AND GEOMETRY, AND FELLOW OF
ST JOHN'S COLLEGE, IN THE UNIVERSITY OF CAMBRIDGE

VOLUME I
FOUNDATIONS

Amplissima est et pulcherrima
scientia figurarum

CAMBRIDGE
AT THE UNIVERSITY PRESS
1922

PREFACE

THE volumes of which this is the first have the purpose of introducing the reader to those parts of geometry which precede the theory of higher plane curves and of irrational surfaces. The present volume is devoted to the indispensable logical preliminaries. It assumes only those relations of position, for points, lines and planes, which, furnished with a pencil, a ruler, some rods and some string, a student may learn by drawing diagrams and making models. It seeks to set these relations in an ordered framework of deduction, gradually rendered comprehensive and precise enough to include all the subsequent theory; to this end it puts aside, at first, most of those intricate details which make up the burden of what is generally called elementary geometry. That such a plan can be carried through, thanks to the work of many generations of thinkers, is well enough known; and experience has shewn that many students, especially of the class who look forward to becoming Engineers or Physicists, to whom the geometry of the usual text-books is tiresome, find such a course stimulating and easy, when the matter is properly presented to them. The mathematician who has followed such a course will find that he has no cause to think he has learnt the wrong things. The fundamental theorems in this method of approaching the subject are indeed of Greek origin; only, these are here made to lead to general principles, giving a command of detail unknown to the Greeks. Subsequent volumes will deal, on the basis of the results obtained in this volume, with conics (and circles), with quadric surfaces and cubic curves in space, and with cubic surfaces and certain quartic surfaces. These volumes are ready to print; it is hoped that they may appear in no long time.

Speaking in more detail of the present volume, it rejects the consideration of distance, and of congruence, as fundamental ideas; these are, in effect, replaced by a theory of related ranges; the geometry usually described with the help of the notion of distance appears later, in a more general, but not more difficult, form. By what means it is possible, so to dispense with this notion, should be of interest to others than the student of geometry. An account is given, however, of the consequences of accepting as fundamental the continuity of the real points of the line. As it is necessary to provide for the consideration of the so-called imaginary elements, room has been given to the justification of these; the ideas to which they lead are indeed an essential part of the power which belongs to the point of view adopted. The references to the theory of space of more than three dimensions are also vital to the plan of the work; it will be seen later with what simplicity this theory enables us to deal with curves and surfaces whose properties can otherwise be developed only at great length, with complicated analysis. The geometrical theory is accompanied by an algebraic symbolism, which serves to help to fix ideas, and for purposes of verification; it is necessary also to include the proof that this symbolism is appropriate to the purpose. It is held, however, that the geometrical argument should be complete in itself, independently of the symbols; a geometry should have such a comprehensive grasp of geometrical relations that all its results are clear by consideration of the geometrical entities alone.

The writer is under great obligations to Mr J. B. Peace, M.A., for his interest in the work, and to the Staff of the University Press for their ready cooperation, especially at this time of difficulty in the production of books.

H. F. B.

26 *September* 1921

TABLE OF CONTENTS

Section III. Introduction of algebraic symbols

CHAPTER II.   REAL GEOMETRY

Section I. The Propositions of Incidence. Introduction of a plane, and of a space

Section II. The extension of the Real Geometry by means of Postulated Elements

Contents

*Section III.   A deduction of Pappus' theorem from the
extended Real Geometry*

CHAPTER III.   ABSTRACT GEOMETRY, RESUMED

*Related spaces; justification of the symbols.   Geometrical assumption
for imaginary elements; replacement of imaginary elements by
series of real elements*

INTRODUCTORY

THE present is the first volume of a brief attempt to place the reader in touch with the main ideas dominant in contemporary geometry, excluding the consideration of higher algebraic irrationalities. For this purpose an account of many of the preliminary facts of geometry is indispensable, but in later stages a good deal of variety was possible in the selection, and no complete recital of geometrical theorems is aimed at. In several respects the views taken are not those usual in current textbooks; but it is believed that the system here suggested is logically complete, and does not require that long preliminary study of elementary geometry to which at present so much time is devoted. In particular it is desired to enter a protest against the custom of regarding so-called projective geometry as based upon metrical geometry; in the present account *distance*, as a primary conception, does not enter at all. And an attempt is made to include as soon as possible the indispensable ideas of geometry of more than three dimensions, and of geometry of so-called imaginary points. While the view is taken that all geometrical deduction should finally be synthetic, it is also held that to exclude algebraic symbolism would be analogous to preventing a physicist from testing his theories by experiment; and it becomes part of the task to justify the use of this symbolism.

It was impossible on such a plan to avoid discussing the logical foundations—and to this the present volume is devoted. Here difficulties of exposition arise from the desire to be brief, and not to be pedantic. It is not so much aimed at to give a faithful transcript of views elsewhere stated to be final, as to suggest a position which may appeal to the reader's logical sense as likely to be satisfying with further analysis. Logically, Chapter II, headed *Real Geometry*, should have come first, and should have been developed in greater detail. To the writer, after much experience as a teacher, that seemed unpractical. He has thought it better to state at once, in Chapter I, rather as working hypotheses, the greater part of the foundations on which the whole is to be based, reserving however, for Chapter III, an account of the grounds on which imaginary elements are introduced, as well as a more formal *résumé* of the logical position adopted. Thus Chapter II is a less abstract, but so far as the writer can judge the reader's view, a more toilsome analysis of fundamental conceptions than Chapter I. Incidentally, the particular geometrical propositions of Chapter I furnish a good deal of practice in the habit of geometrical construction in three dimensions, which, to the writer, appears logically prior to a detailed consideration of plane figures.

A Science grows up from the desire to bring the results of observation, of the relations of a class of facts which appear to be connected, under as few general propositions as possible. Into these propositions it is generally found necessary, or convenient, when the science has reached a sufficient development, to introduce abstract entities, transcending actual observation, whose existence is only asserted by the postulation of their mutual relations. If the science is to be arranged as a body of thought developed deductively, it is necessary to begin by formulating fundamental relations connecting *all* the entities which are to be discussed, from which other properties are to follow as a logical consequence. If this is done we may in the first instance regard all the entities involved in these fundamental propositions as being abstract, even those to which we attach names agreeing with the names borne by entities which we regard as subject to actual observation. The usefulness of the science, for the purpose for which it was undertaken, will depend on the agreement of the relations obtained for these latter entities with those which we can observe. It would seem that this process of substituting conceived entities, limited by supposed interrelations, for those which are regarded as objects of experience, belongs to every science. But it is clear that the degree of abstractness which may usefully and safely be applied is a matter for judgment and choice, conditioned by knowledge of the matter in hand, even in dealing with the same experiences. In geometry, as applied to the external world, we cannot but be conscious, for instance, in dealing with the points of a line, of the difference between those points which we regard as accessible and those which we regard as the others; and then, given two points of the line, of the difference between those points which we think of as between these, and those not between these; and then, finally, of the order in which several points of the line, which we can think of simultaneously, are arranged. None of these considerations however is taken account of in Chapter I; nor indeed do they enter finally into the abstract system of geometry which we set up. But it is to be expected that, if we take account of them, we may be able to analyse further some of the conceptions which we have adopted without analysis in Chapter I. It is to shew how this can be carried through that Chapter II is written. In particular this Chapter gives grounds on which the fundamental theorem for the correspondence of points on two straight lines may be based. The theory is built up in this Chapter with the use only of points supposed accessible and lying in a limited (unclosed) region, and there are some striking differences between the propositions and those of Chapter I; for instance it is not assumed that two lines in a plane necessarily have a point of intersection; and a line in a plane divides the points of the plane into two distinct sets. It is therefore of importance to shew that there is no inconsistency in afterwards disregarding these special results, as we do; this is effected by shewing that we may consider the limited space as a portion of a closed space in which the more general, if less definite, propositions of Chapter I are true. This step is of interest too as shewing how the detailed considerations which make up the usual elementary expositions of the so-called non-Euclidian geometry are superseded by the more general point of view. To some extent the comparison between the two points of view is analogous to the comparison between the theory of quadratic equations in which there may or may not be two roots, and the theory in which there are always two roots.

The writer is not of those who hold that the process of analysing the fundamental conceptions and setting out the sufficient axioms can ever be final; any such formulation, it would seem, must be subject to the possibility that other logical alternatives of consecutive thought may be revealed. Indeed the progress of science appears to consist in the very gradual unfolding of such alternative, or more embracing, conceptions. The history of Mathematics furnishes many examples, none more instructive and interesting than in the case of non-Euclidian geometry. But the instinct to such analysis is not less imperative than the instinct to all scientific thinking, and in geometry as in other subjects the attempt has been fruitful in a better understanding of the subject-matter. From those who would base upon the confession of the incompleteness of the analysis a refusal to enter upon it the writer would appeal to this instinct, which is more fundamental than formal logic.

Such a volume as this would be impossible save for the work and insight of a long line of other writers. In the Bibliography at the end an attempt is made to give the most necessary acknowledgments. In the text the view is sometimes taken that, in Mathematics, an indication may be more acceptable than an exposition; one remembers the words of Desargues (1636, *Oeuvres* I, 420) " la démonstration est...en notes sans discours pour le soulagement de la veuë et de l'esprit."

CHAPTER I

ABSTRACT GEOMETRY

SECTION I. PROPOSITIONS OF INCIDENCE

Propositions of incidence in three dimensions. The entities with which we deal in the first instance we call by the names, *point*, *line*, and *plane*. These are any objects which are subject to the following laws of combination, which we call the Propositions of Incidence, together with another law, explained below (Sect. II). It is provisionally assumed that these laws are self-consistent and, when properly explained, are sufficient to enable the reader to form a clear impression whether any statement made in regard to these entities is a consequence, or not, of the fundamental laws. It is also very frequently assumed provisionally, when, in the course of a geometrical construction, two points are obtained by certain rules and it is desired to continue the construction with the help of the line joining these points, that these points do not coincide. In many cases it may be easy to shew that the coincidence of the points would involve an undesired limitation in the given points of the figure. But there may be other cases in which it would be consistent with the assumed fundamental propositions to assume either that the two points always coincide, or that they do not always coincide. In taking the latter alternative we should then be neglecting possibilities which, even if special, may quite well be worth examination. We adopt words which are in common use, but it is not to be assumed that in effect the meanings which we attach to these are co-extensive with the usual meanings. As is customary we frequently employ diagrams to fix ideas and clarify statements, representing a line by a mark made on the paper and a point by a dot thereon. Notwithstanding this formal and abstract method of statement, the whole object of the theory is of course, as has been stated, to deal with the ideas suggested by our ordinary experience; *it is this experience itself which has gradually suggested the abstract statement.* The Propositions may be stated as follows:

Through an arbitrary point there pass an infinite number of lines, of which one passes through any other arbitrary point. Thus a line is determined uniquely by any two points. The line contains an infinite number of points beside the two determining ones, and is determined by any two of these. Through any given line there pass an infinite number of planes, of which one passes through any point not lying on the given line. Thus a plane is determined uniquely by any three points which do not lie in line. The plane contains an infinite number of points beside the three determining ones, and is determined by any three of these which do not lie in line. The plane entirely contains the line which is determined by any two of its points. Thus if two lines have a common point, and we take two other points, one on each of the lines, the plane determined by the three points contains both the lines; that is, two lines with a common point determine a plane, containing both the lines. It is also true that two lines which are in the same plane have a common point. We have said that there is an infinite number of planes all containing the points of any given line; it is also true that any two planes have common points, lying on a line, or that any two planes intersect in a line. It will be seen below that this statement ceases to be true when we are not limiting ourselves to space of three dimensions. Further it is true (still in space of three dimensions), that any plane contains a point of any line (and contains all the points of the line if it contains two of these), or that a line and a plane intersect in one point. Thus any three planes, which have not in common all the points of a line, have a point in common, that namely where the line of intersection of two of the planes meets the third plane. If now we compare this with the earlier statement that three points not in line determine a plane, and compare the other statements with one another in pairs, as for instance the statement that two points determine a line with the statement that two planes determine a line, the statement that a line and a point determine a plane with the statement that a line and a plane have a point in common, the statement that two lines in a plane have a point in common with the statement that two lines intersecting in a point lie in a plane, we see that there is a complete descriptive correspondence between the relations of points, lines and planes on the one hand, and the relations of planes, lines and points on the other hand. Two propositions obtained one from the other by this interchange will be said to be *reciprocal*, or *dual*, or *correlative*. If we limit our consideration to the points and lines of a single plane it is at once seen that there is a similar reciprocity between propositions relating to points and lines and other propositions relating to lines and points. And a similar duality will be found to subsist in any space of higher dimensions.

We now at once deduce several immediate consequences of these Propositions of Incidence.

The transversal to two lines from a point. Given any two lines a, b, which do not intersect, or, as we shall often say, are *skew* to one another, and a point, P, which does not lie on either of these lines, a single line can be drawn from P to meet both a and b. For

P and *a* determine a plane, as do *P* and *b*; and *P* is among the points common to these planes, that is, it is on their line of intersection. But this line, being on the plane *Pa*, intersects the line *a*, and, being on the plane *Pb*, intersects the line *b*; it is thus such a line as is desired. Conversely, any line through *P* which meets the line *a* and *b* must be on both the planes *Pa* and *Pb*, and must therefore be the line we have found. The correlative proposition is that there lies in any plane not containing either of two given skew lines a line which meets both; the join namely of the points where the plane intersects the respective lines.

Desargues' theorem. If two triads of points, *A*, *B*, *C* and *A'*, *B'*, *C'*, be such that the three joining lines *AA'*, *BB'*, *CC'* meet in one point, say *O*, then the two lines *BC*, *B'C'* meet, say in the point *P*, and similarly the lines *CA*, *C'A'* meet, say in *Q*, and the lines *AB*, *A'B'* meet, say in *R*, and the three points *P*, *Q*, *R* are in line. Conversely, given two triads of points, *A*, *B*, *C* and *A'*, *B'*, *C'*, such that the lines *BC*, *B'C'* intersect, say in *P*, that the lines *CA*, *C'A'* intersect, say in *Q*, and that the lines *AB*, *A'B'* intersect, say in *R*, while *P*, *Q*, *R* are in line, then the lines *AA'*, *BB'*, *CC'* cointersect.

It is supposed that the points *A*, *B*, *C* are not in line, so that they determine a plane, and similarly that *A'*, *B'*, *C'* determine a plane. Take first the case when these planes are different. Suppose further that no one of the joins *BC*, *CA*, *AB* is in the plane *A'B'C'*, and no one of the joins *B'C'*, *C'A'*, *A'B'* is in the plane *ABC*. Then the fact that the lines *BB'*, *CC'* intersect involves that the points *B*, *B'*, *C*, *C'* are in one plane, and hence that the lines *BC*, *B'C'* meet. Their point of intersection, say *P*, lies then on *BC* which is in the plane *ABC*, and on *B'C'* which is in the plane *A'B'C'*, so that *P* lies on the line of intersection of these two planes. The point, *Q*, of intersection of the lines *CA*, *C'A'*, similarly found, as well as the point, *R*, of intersection of the lines *AB*, *A'B'*, are equally on the line of intersection of the planes *ABC*, *A'B'C'*. So that the former statement made is clearly true. Conversely if the lines *BC*, *B'C'* meet in the point *P*, while *CA*, *C'A'* meet in *Q* and *AB*, *A'B'* meet in *R*, then, still assuming the planes *ABC*, *A'B'C'* to be different, and assuming that no one of the lines *BC*, *B'C'*, *CA*, *C'A'*, *AB*, *A'B'* is common to both planes, the points *P*, *Q*, *R* are evidently in the line of intersection of these planes. The fact that the lines *BC*, *B'C'* intersect, involves that the lines *BB'* and *CC'* intersect, while similarly the line *AA'* intersects both of these. Now the lines *BB'*, *CC'* determine a plane, and the line *AA'* does not lie in this plane, since we have assumed that the planes *ABC*, *A'B'C'* are different; wherefore the line *AA'* meets this plane in a point and can only intersect both *BB'* and *CC'* by passing through their point of intersection. Thus the lines AA', BB', CC' meet in a point, as we desired to shew.

We have explicitly excluded, for the sake of simplicity, the possibility of one of the six lines BC, ... lying in both planes, which is easily seen to be unimportant. Incidentally we see that if three lines be such that every two of them have a common point, then the three lines lie all in one plane, or pass through one point.

Next suppose the two triads A, B, C and A', B', C' to be in one plane, it being assumed as before that A, B, C are not in line and A', B', C' are not in line. Then, first, let the lines AA', BB', CC' meet in one point, O. Draw through O any line not lying in the plane of the two triads, and let P, P' be any two points on this line. Then the intersecting lines $AA'O$ and $PP'O$ determine a plane, and the lines AP, $A'P'$, lying therein, intersect in a point, say A''. Similarly the lines BP, $B'P'$ meet in a point, say B'', and the lines CP, $C'P'$ meet in a point, say C''. As P is not in the plane of ABC, $A'B'C'$, it is clear that the points A'', B'', C'' are not in this plane; further A'', B'', C'' are not in line, since otherwise, as the plane containing this line and P would contain the points A, B, C, these would be in line; finally the plane which is thus determined by the points A'', B'', C'' does not contain, for instance, the line AB, for otherwise the plane of AB and $A''B''$, which contains P, would contain C'' and C, and A, B, C would be in line. Hence it follows from what is proved above, as the lines AA'', BB'', CC'' meet in P, that there is, in the plane of ABC and $A'B'C'$, a line, where this plane is met by the plane $A''B''C''$, and upon this line three points, say L, M and N, such that the lines BC, $B''C''$ meet in L, the lines CA, $C''A''$ meet in M and the lines AB, $A''B''$ meet in N. The point L is then the point where the plane of ABC and $A'B'C'$ is met by the line $B''C''$, and M, N are similarly the points where this plane is met by $C''A''$ and $A''B''$. By considering that the three lines $A'A''$, $B'B''$, $C'C''$ meet in the point P', we prove however, in the same way, that $B'C'$ passes through L, and that $C'A'$, $A'B'$ respectively pass through M and N. We have thus proved that if two triads of points ABC, $A'B'C'$, in the same plane, be such that the joining lines AA', BB', CC' meet in a point, then the three points of intersection $(BC, B'C')$, $(CA, C'A')$, $(AB, A'B')$ are in line. Conversely, for two triads of points ABC, $A'B'C'$ in the same plane, assume that the three points $(BC, B'C')$, $(CA, C'A)$, $(AB, A'B')$ are in line. Denote these points respectively by L, M, N. Draw through the line LMN a plane, other than the original plane of ABC, $A'B'C'$; draw through the points L, M, N, in this new plane, respectively the lines $LB''C''$, $MC''A''$, $NA''B''$, giving by their intersections a further triad A'', B'', C'' in this new plane. This new plane does not contain any one of the lines BC, CA, AB, $B'C'$,

$C'A'$, $A'B'$, and it is supposed that the lines $LB''C''$, $MC''A''$, $NA''B''$ are all drawn so as not to lie in the original plane of ABC, $A'B'C'$. Then, as the lines BC, $B''C''$ meet in the point L, the lines CA, $C''A''$ meet in the point M, and the lines AB, $A''B''$ meet in the point N, it follows from what has preceded that the three lines AA'', BB'', CC'' meet in a point, say P, which does not lie in either the original plane or the new plane. It follows similarly, as the lines $B'C'$, $B''C''$ meet in the point L, and $C'A'$, $C''A''$ in M, and $A'B'$, $A''B''$ in N, that the lines $A'A''$, $B'B''$, $C'C''$ meet in a point, say P', not lying in either of the two planes described. Now let O be the point where the line PP' meets the original plane of ABC, $A'B'C'$. Then, since the lines AP, $A'P'$ intersect, in the point A'', and the points A, A', P, P' are not in line, these lines determine a plane, and therefore the line AA' intersects the line PP'. This intersection can only be at O, where the line PP' meets the original plane containing A and A'. Thus the line AA' passes through O, as, similarly, do also the lines BB' and CC'. We have thus proved that if two triads ABC, $A'B'C'$, of points in one plane, neither triad being in line, be such as to have three intersections $(BC, B'C')$, $(CA, C'A')$, $(AB, A'B')$ lying in line, then the three joining lines AA', BB', CC' meet in a point.

The aggregate of the four theorems now established will be referred to as Desargues' theorem. (See *Oeuvres de Desargues, par M. Poudra*, Paris, 1864, t. I, pp. 413, 430.)

Remarks in regard to Desargues' theorem. (*a*) The reader will notice that the theorem for two triads of points in one plane is proved by considerations involving the existence of points not lying in this plane. From the previously given Propositions of Incidence it is possible to select those which relate only to the geometry of points and lines lying in one plane. It is an interesting fact that the theorem of Desargues for two triads of points lying in this plane cannot be proved as a consequence only of the propositions of incidence relating to geometry in this plane—as we shall prove below (Chap. II). The propositions of incidence for one plane appear to furnish no criterion for deciding whether three constructed points lie in line; such criterion arises only when the line enters as the intersection of two planes. And they appear to furnish no criterion for deciding whether three constructed lines meet in one point; such criterion arises only when the point enters as the intersection of the plane with a line not lying therein.

(*b*) We have said that two lines intersect one another when they lie in one plane; thus they intersect one another when there are two points A, A' on one of the lines, a, and two points B, B' on the other line, b, such that the cross joins AB, $A'B'$ meet; for then the two lines a, b lie in the plane of these lines AB, $A'B'$. Let

F be the intersection of AB and $A'B'$; and let C be any other point not lying on either of the lines a, b. Desargues' theorem furnishes a construction for the line joining C to the point of intersection of the lines, a, b; and this construction depends only on the points A, A' and B, B' by which these lines may be supposed to be determined.

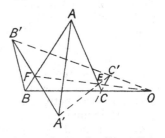

For, on the given line BC, in the plane ABC, wherein A is also given, take a point O and join it to the given point F of the line AB. Let the joining line meet AC in the point E. The plane determined by the points O, B', F contains the points A' on $B'F$, and E on FO; thus the line $A'E$ intersects OB', say in C'. We now have two triads of points, ABC and $A'B'C'$, such that the lines BC, $B'C'$ meet in O, the lines CA, $C'A'$ meet in E, and the lines AB, $A'B'$ meet in F, while the points O, E, F are in line. It therefore follows that the line CC' passes through the intersection of the given lines AA' and BB'.

That this construction should be useful when the point of intersection of the lines AA', BB' is not accessible depends on the possibility of taking O in such a position upon BC that the two points E, C' should be accessible (see Chapter II). The construction is valid when C is in the plane of the two lines AA', BB'; when C is not in this plane the line from C to the intersection of these lines may also be defined as the line of intersection of the two planes CAA' and CBB'.

(*c*) If we have two tetrads of points, A, B, C, D and A', B', C', D', with the property that the four lines AA', BB', CC', DD' meet in one point O, it being supposed that neither tetrad consists of four points lying on a plane, then it can be shewn that every two corresponding joining lines in the two figures, such as BC and $B'C'$, intersect one another, and that the six points of intersection so arising are in one plane, and are the intersections in pairs of four lines of that plane.

In fact, the lines DA, $D'A'$, in the plane of the lines ODD', OAA', must intersect, say in P; so the lines DB, $D'B'$ must intersect, say in Q, and the lines DC, $D'C'$ must intersect, say in R. The points P, Q, R are then not in line, since the points D, A, B, C are not in a plane; and thus the points P, Q, R define a plane. Again, as the joins of corresponding points of the two triads DBC and $D'B'C'$, namely DD', BB' and CC', meet, in O, it follows from Desargues' theorem that the lines BC, $B'C'$ intersect in a point of the line QR, where Q is the point $(DB, D'B')$ and R is the point $(DC, D'C')$, say in X. Similarly the lines CA, $C'A'$ meet in a point, say Y, of the line *RP*, and *AB*, *A′B′* meet in a point, say *Z*, of the line *PQ*. The points *X*, *Y*, *Z* are on the plane *ABC*, and are thus in line, the line of intersection of the planes *ABC* and *PQR*. This proves the result stated.

(*d*) Two figures, such as the triads *ABC*, *A′B′C′* of Desargues' theorem, or the two tetrads just considered, are said *to be in perspective*, when to any point *P* of one figure there corresponds a point *P′* of the other figure, such that all lines, like *PP′*, which join two corresponding points, pass through the same point.

We may notice that the figure which arises in the proof of Desargues' theorem for the case of two triads in one plane, consists of three triads, of which every two are in perspective, the three centres of perspective being in line; while each set of three corresponding joining lines, such as *BC*, *B′C′*, *B″C‴*, is formed of three lines which meet in a point. In all there are fifteen points and twenty lines; each line contains three of the points, and through each point there pass four of the lines. If we denote the points *P*, *P′*, *A″*, *B″*, *C″* respectively by the binary symbols, each formed with two numbers, 04, 05, 01, 02, 03, then it is appropriate to denote the points *A*, *B*, *C*, *A′*, *B′*, *C′* respectively by the symbols 14, 24, 34, 15, 25, 35, the point of meeting of the three lines *BC*, *B′C′*, *B″C″* by 23, the point of meeting of the three lines *CA*, *C′A′*, *C″A″* by 31, and that of *AB*, *A′B′*, *A″B″* by 12, using 45 for the point of meeting of *AA′*, *BB′*, *CC′*. The points are then denoted by all the combinations of six symbols two at a time. Each line is met by nine others, lying in threes in three planes; the three triads so formed, one in each of these planes, are two and two in perspective, the centres of perspective being in line.

The figure arising in the proof of Desargues' theorem for two triads not lying in the same plane is simpler. Here there are five planes, ten lines, consisting of the lines of intersection of the planes in pairs, and ten points, consisting of the intersections of the planes in threes.

Example. If two triads of points, *A*, *B*, *C* and *A′*, *B′*, *C′*, be such that the lines *AB′*, *A′B* intersect, say in *R*, also *AC′*, *A′C* intersect, say in *Q*, while *AB*, *A′B′* intersect in *H*, and *BC*, *B′C′* intersect, in *F*, then prove that *BC′*, *B′C* intersect, say in *P*, and that *AC*, *A′C′* intersect, say in *G*, and that each of the four sets *FGH*, *FQR*, *GRP*, *HPQ* consists of three points in line.

The fourth harmonic point. We come now to an important construction by which we pass from three given points of a line to another unique point of this line. For the construction itself it is assumed that a plane can be drawn through the line, and for the proof that the fourth point of the line which is constructed is unique it is assumed that two planes can be drawn through the line. Thus the construction supposes that the line lies in space of three dimensions, though the result of the construction relates only to points lying on a line.

Let A, B, C be three arbitrary points lying on a line. Through this line draw an arbitrary plane. In this plane draw three lines, one through each of the points A, B, C, but otherwise arbitrary, forming by their intersections a triad P, Q, R, so that the line QR passes through the point A, while RP passes through B and PQ passes through C. *Now regard A, B, C as consisting of a pair of points, and a third point*; let A, B be the pair. Join to A the point of intersection of the line (RP) drawn through B with the line (PQ) drawn through C; that is, join AP. Similarly join to B the point of intersection of the line (QR) drawn through A with the line (PQ) drawn through C; that is, join BQ. Let the lines AP and BQ meet in the point U. Join U to the point of intersection of the line (QR) drawn through A with the line (RP) drawn through B; that is, join UR. Let UR meet AB in D. Then D is the point to be constructed corresponding to the separation of the points A, B, C into the pair A, B and the third point C. Temporarily it may be denoted by

$$D = (A, B)/C.$$

We may similarly construct two other points, appropriately represented respectively by

$$(B, C)/A \text{ and } (C, A)/B,$$

with which we are not concerned for the moment.

To prove that the point D is the same whatever be the lines QR, RP, PQ we proceed as follows. Draw through the given line another plane, and therein make a similar construction, with lines $Q'R', R'P', P'Q'$ passing respectively through A, B and C. We shew first that the points D, D' in which $RU, R'U'$ meet AB are the same. In fact the planes RAR', RBR', are intersected by the plane $QPCP'Q'$ respectively in the lines QQ' and PP'; these lines, lying in a plane, must then meet, and their

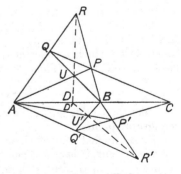

point of meeting must be on the line RR' containing all the points common to the planes RAR', RBR'. Let this common point of the lines QQ', PP' be called O. Again the planes PAP', QBQ' are met by the plane $QQ'OP'P$ respectively in the lines PP' and QQ';

these lines, lying in a plane, must then meet in a point on the line of intersection UU' of the planes PAP', QBQ'.

Thus the line UU' passes through the intersection, O, of the lines PP' and QQ', through which the line RR' also passes. But, from the fact that UU' and RR' meet, it follows that the lines RU and $R'U'$ lie in a plane, so that these intersect; and their intersection, being common to the planes RAB, $R'A'B'$ in which they respectively lie, must be on the line AB.

Next suppose that in the plane PQR other lines are drawn through A, B, C, namely AQ_1R_1 through A, and BR_1P_1 through B, and CP_1Q_1 through C; and let AP_1, BQ_1 meet in U_1, while R_1U_1 meets AB in D_1. By the argument just given it follows that D_1 coincides with the point, D', in which $R'U'$ meets AB. And this has been shewn to coincide with the point, D, in which RU meets AB.

The position of D is thus independent of the lines AQR, BRP, CPQ.

First properties of the harmonic relation. (*a*) It is clear from the construction that no change is made in D by the interchange of the points A and B. Thus we may equally write

$$D = (B, A)/C.$$

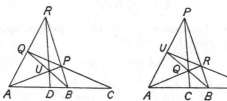

(*b*) Again, as we may see using the same figure but calling the points originally named P, Q, R, U, C, D respectively by the names R, U, P, Q, D, C, we can pass from the point D to the point C by a construction which is descriptively identical with that by which we passed from C to D. Thus we may also write

$$C = (A, B)/D,$$

and, like the couple A, B, the points C, D form a couple of interchangeable points.

(*c*) And in fact one of these couples may be interchanged with the other. For, it is clear from the original construction that if U_1 be any point on the line RD, and the lines AU_1, BU_1 meet RB and RA respectively in P_1 and Q_1, then P_1Q_1 passes through C; we have only to replace the line CPR by the line CQ_1; if the intersection of this with RB be joined to A, the joining line must meet BQ_1 in a point lying on RD, that is, in U_1. Hence, interchanging the parts played by the lines CBA and CPQ, if the line PD meet RA in Q', and the line QD meet RB in P', it follows from the fact that the lines PA, QB meet on AD that

the line $Q'P'$ passes through C.

Now consider the point which would be expressed by

$$(D, C)/B.$$

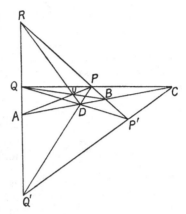

To obtain this point by the rule, we may draw respectively through the points D and C the lines DP' and CP, and then through the point B the line $P'BP$, meeting the two former respectively in the points P' and P; then, if the lines $P'C$ and PD meet in Q', we are to join Q' to the point of intersection of the lines DP' and CP, that is to the point Q. The joining line QQ' will then meet DC in the point required; and this is the point A. Hence we have

$$A = (D, C)/B,$$

which shews that the couples A, B and C, D may be interchanged.

We may then speak of *the points of either couple as being harmonic conjugates of one another in regard to the points of the other couple.*

(*d*) We have stated that two figures are said to be in perspective when to any point, P, of one figure there corresponds a point, P', of the other, such that the joining lines, PP', all pass through the same point. It is easy to see that if four points A', D', B', C' of one line be in perspective with four points A, D, B, C respectively, of another line, which are such that C, D are harmonic conjugates in regard to A, B, then also C', D' are harmonic conjugates in regard to A', B'.

If, in the original figure, the line RU meet the line CPQ in the point X, then X is the harmonic conjugate of C in regard to P and Q. For, from P, Q are drawn, respectively, the lines PA, QB, meeting in U, and the line CAB, drawn from C, meets these respectively in A and B; then the lines AQ, BP meet in the point R, and the line RU meets the line PQ in the point X.

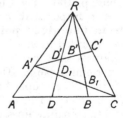

Hence taking the points A', D', B', C' respectively in perspective with the points A, D, B, C, from the centre R, where C, D are harmonic conjugates in regard to A and B, let the line CA' meet the lines RD and RB respectively in D_1 and B_1. From the remark just made, it follows first that C, D_1 are harmonic conjugates in regard to A' and B_1; and then it similarly follows that C', D' are harmonic conjugates in regard to A' and B'.

(*e*) Conversely if on each of two lines we have a set of four points in harmonic relation, they can be regarded as both arising by perspective from the same set of four points in harmonic relation lying on another line.

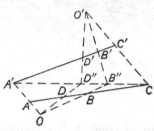

For let C, D be harmonic conjugates in regard to A, B, on one line, and C', D' be harmonic conjugates in regard to A', B', on another line; join $A'C$ and take O any point on $A'A$; let OD, OB meet $A'C$ in D'', B'' respectively; let CC' and $B''B'$ meet in O' and let $O'D'$ meet $A'C$ in D_1. Then, by what is remarked under (*d*), C and D'' are harmonic conjugates in regard to A' and B'', while C and D_1 are also harmonic conjugates in regard to A' and B''. Thus D_1 is the same as D'', and the two given sets are both in perspective with the harmonic set A', D'', B'', C.

(*f*) When the points that can exist on a line are further conditioned as in Chapter II below, it will be found that if the points C, D are harmonic conjugates in regard to the points A, B, then C and D are necessarily separated from one another, in the sense there explained, by A and B. Without these conditions, to speak of a separation would be unmeaning. But we shall, in accordance with a general assumption referred to at the beginning (above, p. 4), assume provisionally that the point D does not coincide with C unless C coincides with A or with B. A geometry, of fifteen points in all, has been suggested in which D always does coincide with C (Fano, "Sui postulati fondamentali...," *Giorn. di Mat.*, xxx, 1892, p. 106). See Chap. II, Sect. II, below.

An important construction determining a sixth point from five given points of a line. The five given points are to be regarded as consisting of two couples and another point; for clearness sake we shall at present speak of this last point as the determining point. Denote the points of one couple by A, B, those of the other couple by O, U; let E be the determining point, and P the point to be determined. Draw any plane through the line, and therein draw lines, denoted respectively by a, b, o, u, one through each of the points A, B, O, U of the two given couples. We associate one of the two lines a, b with one of the two lines o, u, and the other of a, b with the remaining one of o, u; this is possible in two ways; take the association of a with u and of b with o, denoting the intersection of a and u by (a, u), and the intersection of b and o by (b, o). The lines a, b, o, u are to be any lines of the plane taken subject to the condition that the two points (a, u), (b, o) shall lie on a line, say e, which passes through the determining point E. Then by the other possibility of association we have two other points which may be respectively denoted by (a, o) and (b, u). The line, say p, which joins these, gives, by its intersection with the original line, the point, P, which is determined, in the one of two possible ways chosen, as the result of the construction. The other method of association would require the lines a, b, o, u to be such that the points (a, o) and (b, u) shall lie on a line, e', passing through E, and then the line, p', containing the points (a, u) and (b, o), would meet the original line in a point, P', which would be the point determined by the construction. For our purpose it will be found to be very important that the Propositions of Incidence assumed do not enable us to prove that P and P' coincide.

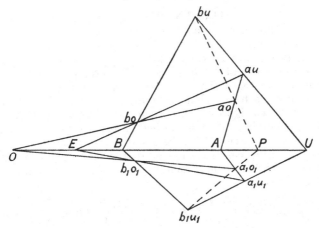

That, subject to the condition that the points (a, u), (b, o) shall lie on a line through E, the point P is the same whatever be the lines a, b, o, u, may be proved as in a previous case.

Draw through the original line any other plane, and therein, through the points A, B, O, U, respectively, any lines a_1, b_1, o_1, u_1, subject to the condition that the points (a_1, u_1) and (b_1, o_1) shall lie on a line through E, completing the figure as before. We shew that the line joining the points (a_1, o_1) and (b_1, u_1) also passes through P. For, first, the three lines joining the respective pairs of points

(a, u) and (a_1, u_1), (b, o) and (b_1, o_1), (a, o) and (a_1, o_1)

meet in a point, namely in the point of intersection of the three planes

$$[e, e_1], \quad [a, a_1], \quad [o, o_1],$$

where e, e_1 are the lines through E in the two planes respectively; this follows because each of these planes contains two of the three joining lines in question, or each of the lines is the line of intersection of two of these planes. Similarly, the three lines joining the respective pairs of points

$$(b, o) \text{ and } (b_1, o_1), \quad (a, u) \text{ and } (a_1, u_1), \quad (b, u) \text{ and } (b_1, u_1)$$

meet in a point, namely in the point of intersection of the three planes

$$[e, e_1], \quad [b, b_1], \quad [u, u_1],$$

each of these lines being common to two of these planes.

From these two facts it follows that the two lines joining the respective pairs of points

$$(a, o) \text{ and } (a_1, o_1), \quad (b, u) \text{ and } (b_1, u_1)$$

must intersect one another. From this it follows that the two lines joining the respective pairs of points

$$(a, o) \text{ and } (b, u), \quad (a_1, o_1) \text{ and } (b_1, u_1)$$

must meet; and they can only meet in a point P of the original line which is the line of all the points common to the two planes in which the two figures are constructed.

Having proved that the same point P is obtained when we construct figures in two different planes passing through the original line, it follows as before that the same point P is obtained when we construct two figures in the same plane, by varying the lines a, b, o, u, preserving the same method of association of the couples a, b and o, u. It is clear that the two couples A, B and O, U play similar parts in the construction, and that the two points of a couple play similar parts; we may therefore represent the points P, P' obtained, respectively, by

$$P = \binom{A, B}{O, U}\Big/ E = \binom{O, U}{A, B}\Big/ E, \quad P' = \binom{A, B}{U, O}\Big/ E = \binom{U, O}{A, B}\Big/ E.$$

Remarks. (*a*) We may make a figure containing both methods of association, using the same lines a, b, u, but two lines o, o' through O.

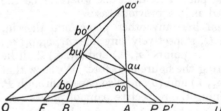

(*b*) As the two methods of association are distinguished by either the interchange of the points A, B, or the interchange of the points O, U, the points P', P will coincide when either A, B coincide, or O, U coincide. In either case the figures appropriate to the two methods of association may be taken identical. When A, B coincide, taking lines a, b, o, u, e for one association, we may take lines $a'=b$, $b'=a$, o, u, e for the other. When O, U coincide, taking lines a, b, o, u, e

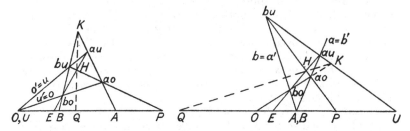

for one association, we may take lines a, b, $o'=u$, $u'=o$, e for the other. By a construction indicated in the diagrams we may prove that when A, B coincide, if $Q=(O,U)/A$, then $P=(A,Q)/E$; and that when O, U coincide, if $Q=(A,B)/O$, then $P=(O,Q)/E$. Hence also, when A, B coincide, if $U=(E,A)/O$, then P coincides with E, as a diagram at once makes clear.

Further, when E coincides with O, both P and P' coincide with U.

A theorem in regard to a chain of perspectivities; reduction to two links. There is another result of great importance to us in the sequel, which may be deduced from the Propositions of Incidence. Though the proof is somewhat intricate we give it at once; the reader may prefer, when he has understood what is the effect of the theorem, to pass over the details of the proof for the present.

It has been seen that two ranges of points $AA'...$, $BB'...$, lying respectively on two lines a, b, are said to be in perspective when the joins of corresponding points of these ranges, that is the lines AB, $A'B'$,..., all pass through another point. It is clearly necessary that the lines a, b should be in one plane. Suppose now that we have a set of lines a, b, c,..., k, finite in number, arranged in order such that every consecutive two intersect; and that there are ranges of points upon these lines, such that those upon the first and second are in perspective with one another from some point, that the ranges upon the second and third lines are in perspective with one another from some point, and so on, for every consecutive two, to the last. *It can then be proved that the ranges upon the first and last lines, a and k, can be obtained from one another by means of only two perspectivities,* namely that there is a line m, and upon this a range which is in perspective with the given range upon the line a from some point, and is also in perspective with the given range upon the line k from some point. It will appear that the line m, which must intersect both a and k, is by no means unique.

Let us agree to say that two ranges of points upon two different lines, p, q, which may or may not intersect one another, are *related*, when these ranges are both in perspective with the same range upon another line, n, from appropriate centres. Two ranges which are in perspective with one another are then also related, in a very particular way, but the converse is not generally true. The proposition to be proved is then that the ranges on the lines a, k are related. And the proposition will follow if we prove the theorem that *when two ranges are related, any range which is in perspective with one of these is related to the other*. We address ourselves then to proving this last theorem.

It may prevent confusion of thought if we add at once that we shall also be led to speak of two ranges of points on the *same* line as being related; but, for a reason which will appear, the definition will be that two such ranges are related when one of them is in perspective with a range on another line which is related, in the above sense, to the other range of the given line, not being directly in perspective with it.

Proof for the case of ranges on skew lines. Before entering upon a general proof of the theorem, we give a proof, which is very direct, for the case when the two given related ranges are on lines which do not intersect one another, and the range which is given as being in perspective with one of the two ranges lies on a line not intersecting the line containing the other of these. Logically it is unnecessary to give this, the general proof which follows being applicable in all cases.

But in this case there is a simple way of describing when the ranges on the two (non-intersecting) lines are related. For if ranges

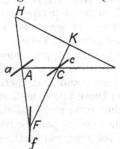

$(A, A', ...)$, $(C, C', ...)$, on two lines a, c, which do not intersect, be both in perspective with a range $(F, F', ...)$, on a line f, so that there is a point H through which the lines $AF, A'F', ...$ all pass, and a point K through which the lines $CF, C'F', ...$ all pass, then the lines $AC, A'C', ...$, joining corresponding points of the two related ranges on a, c, all intersect the line HK. This line HK will not intersect the lines a, or c; for if it intersected a, the plane Ha, which contains f, would contain the line HK, and the plane Kc, containing f and KH, would coincide with the former plane Ha; so that the lines a and c would intersect. Thus a simple construction by which the range on the line a gives rise to the related range on the line *c* is, to draw from the points *A, A',...*, of the range on *a*, the transversals which meet both *c* and the line *HK*; their intersections with the line *c* are the points of the related range thereon. Conversely, if we draw from points *A, A',...* of a line *a*, the transversals to two other lines, *c* and *m*, which do not meet one another, or meet *a*, and if these transversals meet *c* in *C, C',...* respectively, then the ranges (*A, A',...*), (*C, C',...*) are related in the sense first defined; for a range with which they are both in perspective is obtained by taking any two points *H, K* on the line *m*, remarking that the lines *HA, KC* meet in a point, *F*, of the line, *f*, in which the planes *Ha, Kc* intersect one another. We may therefore define two ranges on two non-intersecting lines as being related when the joins of corresponding points of these ranges have a common transversal.

Now consider two related ranges (*A, A', ...*), (*C, C', ...*), on two lines *a, c* which do not intersect one another, the joins *AC, A'C',...* of corresponding points of these ranges having the common transversal *b*, which they meet in (*B, B',...*). Suppose also that the range (*D, D',...*), on the line *d*, is in perspective with the range (*A, A',...*), from the point *O*, the line *d* not intersecting the line *c*. We wish to shew that the range (*D, D',...*) is related to the range (*C, C', ...*).

Let the lines *a, d* meet in *P*; from *P* draw the transversal *PQR* to the lines *b, c*, meeting these respectively in *Q, R*. Let *D, C* be two points, respectively of the ranges on the lines *d, c*, which correspond to a point *A* of the range on *a*; thus *DC* is in the plane joining *O* to *AC*. Wherefore, if *AC* meet *b* in *B*, then *OB* meets *DC*, say in *E*. Next, let *B', D'* be the points of the ranges on the lines *b, d* which similarly arise from *any* point *A'* of the range on *a*, so that *O, A', D'* are in line, and

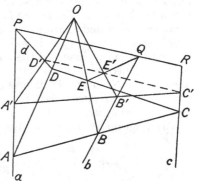

the line *A'B'* meets *c*, in *C'*. The lines *OB', QE* both lie in the plane *Ob*; let them meet in *E'*; we can then shew that *D', E', C'* are in line. For consider the two tetrads of points *A', A, B, B'* and *D', D, E, E'*; as they are in perspective from *O* any two joins of corresponding pairs of points in the two tetrads meet in a point lying on a plane which contains all such points of meeting, as was remarked above in connexion with Desargues' theorem. Such corresponding joins are *AA', DD'*, meeting in *P*, and *BB', EE'*, meeting in Q, and AB, DE, meeting in C; whence it follows that $D'E'$ meets $A'B'$ in a point of the plane PQC, and the only point of $A'B'$ lying on this plane is the point C'. Thus $D'E'$ passes through C'. There- fore, as D' is an arbitrary point of the range on d, we see that this range is related to the range on c, the joins of corresponding points of these ranges all meeting a line QE; this proves what is required. The range thus arising on the line QE is in perspective with the range on the line b, from the point O.

Example. It follows from what has just been given that, if a, b, c be three lines, of which no two intersect, and a range of points on the first be related to a particular range on the second, and this in turn be related to a range on the third, then the range on the first is related to the range on the third; regarding ranges on two non- intersecting lines as being related to one another by the fact that the joins of corresponding points of these ranges have a common transversal, we may speak of this transversal as the *supplementary* line, and the range thereon as the supplementary range. A further exercise in the ideas is then obtained by proving the following particular result: Let a, b, c be three skew (non-intersecting) lines, there being ranges on a and b which are related to one another by a supplementary range on a line x, the range on b being related to a range on c by a supplementary range on a line z. Then if, in accordance with the preceding theorem, the ranges on a and c be related to one another by a supplementary range (y), on a line y, it can be shewn that the ranges (x), (z), which arise on the supple- mentary lines x and z, are also related to one another by the same supplementary range (y). It will be seen that, if the figure formed by four lines in a plane and their six intersections in pairs, or *ver- tices*, be called a complete plane quadrilateral, the result is the same as that, *if of the six vertices of a complete plane quadrilateral, five move on arbitrary skew lines, then the locus of the remaining vertex is also a line.*

The diagram by which we illustrate the proof of this result will

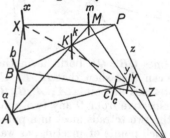 be of a symbolical character, a range being represented by a single general point of the range, except in the case of the range (z). Take any points, P, Q, on the supple- mentary line, z, of the related ranges (b), (c). Then, as has been remarked above, the joins of P, Q, re- spectively to corresponding points, B, C, of the ranges (b), (c), meet in the points of a range (k), lying on the intersection of the planes Pb, Qc. As the ranges (a), (b) are

related, with (x) as supplementary range, and the range (k) is in perspective with the range (b), from P as centre, it follows, from the result just proved, that the joins of corresponding points, A, K, of the ranges $(a), (k)$, meet the plane Px in the points, M, of a range (m), in perspective with (x) from P, which is the supplementary range of the two related ranges (a) and (k). Then, as the ranges $(a), (k)$ are related, with (m) as supplementary range, and the range (c) is in perspective with (k), from Q, it follows, by the same result, that the joins of corresponding points, A, C, of the ranges $(a), (c)$, meet the plane Qm in the points, Y, of a range (y), in perspective with (m) from Q, which is the supplementary range of the ranges $(a), (c)$. These are therefore related. Considering the triads A, B, C and M, P, Q, we see that the joins of corresponding points, AM, BP and CQ, meet in a point K; wherefore, by Desargues' theorem, the three points of intersection of corresponding sides, namely X, from AB and MP, Y from AC and MQ, and Z, from BC and PQ, are in line. Thus any two of the ranges $(x), (y), (z)$ are related, with the third as supplementary range; as we desired to prove. The diagram does not put in evidence the facts that the line k intersects both b and c, and the line m intersects both x and y.

The result is evidently equivalent to saying that if a, b, c, z, x be any five lines of which no two intersect, and from a variable point, B, of the line b, the transversals, ABX, CBZ, be drawn to meet the lines a, x and c, z, respectively in A, X, C, Z, then the lines AC, ZX meet in a point Y lying on a

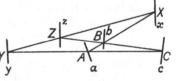

sixth line. This is another form of the statement above made in regard to a complete plane quadrilateral. More generally, given four fixed lines a, c, x, z, of which no two intersect, if through an arbitrary point B we draw the transversal to one pair of these, say a and x, meeting these in A and X respectively, and also draw the transversal to the other pair, meeting these in C and Z, it is clear that the lines AC, XZ will intersect, say in Y (as will also the lines AZ, XC). Taking all the possibilities of grouping there are thereby six points Y obtainable from B, and from any one of these, when the association of the lines is specified, we can proceed back to B by a similar construction.

General proof of the theorem. We now pass to the general proof of the theorem referred to, that, if two ranges be related, any range in perspective with either is related to the other.

By definition two ranges, on different lines, are related when they are both in perspective with a third range, from appropriate centres. For the sake of clearness let this other range be called the intermediary range, the line upon which it lies being called the intermediary line. There is no need to prove the theorem when the two given ranges are in direct perspective with one another, since it follows from the definition of related ranges.

We first shew that the intermediary line, which necessarily in-

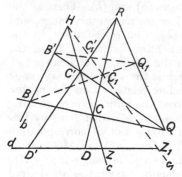

tersects both the given lines, can be taken to be any line whatever intersecting these which does not either (*a*) intersect the lines of the related ranges in corresponding points, or (*b*) pass through the point of intersection of these lines in case they intersect. Let *b*, *d* be the lines of two related ranges, not in perspective with one another, with an intermediary line *c*, intersecting both *b* and *d*. Through the point *H* where *c* intersects *b* draw a line c_1, meeting *d*, in a point, Z_1, not corresponding to the point *H*, or (*b*, *c*), of the range (*b*). The possibility of this requires the assumption that if *b* and *d* intersect, the intermediary line *c* does not pass through their intersection; and this assumption may be made because it is easy to see, from Desargues' theorem, that if the intermediary line *c* passed through the intersection of *b* and *d*, then the ranges on *b* and *d* would be in perspective with one another, which we suppose not to be the case. It is also supposed that c_1 does not coincide with *b* (and therefore does not pass through the intersection of *b* and *d*, if they intersect). We shew that c_1 may be used, instead of *c*, as an intermediary line. From this the general arbitrariness of the intermediary line, asserted above, can be deduced at once. Let the ranges (*b*) and (*c*) be in perspective from the centre *Q*, the points *C*, *C'* of *c* corresponding respectively to the points *B*, *B'* of *b*; also let the ranges (*c*), (*d*) be in perspective from the centre *R*, the points *D*, *D'* of *d* corresponding respectively to the points *C*, *C'* of *c*. The line *QR* may be called the supplementary line. In the plane [*c*, *d*], which contains *R*, and also contains the line c_1 drawn to meet *c* and *d*, let the lines *RC*, *RC'* meet c_1 respectively in C_1 and C_1'. The triads *B*, *C*, C_1 and *B'*, *C'*, C_1' are in perspective from the point *H*; thus, by Desargues' theorem, the lines *BC*, *B'C'* intersect in a point, Q_1, of the line *QR*. Whence, taking for *B'* any point of the range (*b*), with corresponding positions of *C'* and C_1', the range (*b*) is in perspective from Q_1, which is determined by the definite line BC_1, with the range (C_1, C_1', ...) lying on the line c_1, while this is in perspective from *R* with the range (*d*). Thus c_1 is a new intermediary line for the two given ranges (b), (d). The supplementary line QR is unchanged by the replacement of the line c by c_1, being determined however, now, not by R and Q, but by R and Q_1. When we further, in order to get a general replacement of the intermediary line, replace c_1 by a line c', meeting (b), passing through the point (c_1, d), the point R will be replaced by another point, say R', also on QR; the final supplementary line will be the same as the original supplementary line.

For what follows it is necessary to add the following further remark in regard to the choice of the line c_1. Let c and c_1 meet d in Z and Z_1; as the point Q lies in the plane $[b, c]$, the line QZ meets b, say in X, and X, Z are corresponding points of the ranges (b), (d) respectively. Thus X, Z_1 are not corresponding points of the ranges (b), (d). Hence Q_1, Z_1, X are not in line.

Now suppose the ranges (b), (d), on different lines b, d, to be related, not being in perspective with one another, and (a) to be a range in perspective with (b), the line a not being the same as d. We are to prove that the ranges (a), (d) are related. When (b), (d) are in perspective with one another this is true by definition. As before, let (c) be the intermediary range of (b) and (d), the centre of perspective for (b) and (c) being Q. Let X be the point of intersection of the lines a, b, and Z the intersection of the lines c, d. If Q, Z, X are in line, we can, as we have just proved, suppose c and Q respectively replaced by a line c_1 and a point Q_1 such that, if Z_1 be the point (c_1, d), the points Q_1, Z_1, X are not in line. We may, therefore, without loss of generality, suppose Q, Z, X not to be in line. Let h be the line XZ, joining the points (a, b) and (c, d). [See the diagram on page 24.]

Let A, B, C, D be corresponding points of the ranges (a), (b), (c), and (d), respectively, and A', B', C', D' other corresponding points of these ranges. The line h, joining a point of the line b to a point of the line c, is in the plane $[b, c]$, and therefore intersects the lines BC and $B'C'$, say in H and H' respectively. The triads A, B, H and A', B', H' are in perspective from X; thus, if P be the centre of perspective of the ranges (a), (b), lying at the intersection of BA and $B'A'$, it follows, by Desargues' theorem, that $H'A'$ and HA meet on PQ, say in M; this would fail if Q, Z, X were in line, the points H, H' then coinciding at Q. Again the triads H, C, D and H', C', D' are in perspective from Z; if R be the centre of perspective for the ranges (c), (d), it follows, by Desargues' theorem, that $H'D'$ and HD meet on QR, say in N.

This in effect establishes what is required; for the range on (h) is shewn to be in perspective with the range (a), from M as centre, and in perspective with the range (d), from N as centre, so that the ranges (a), (d) are related; the line h is the intermediary line and the line *MN* the supplementary line. The points *M, N* are determined when *A, B, C, D* are assigned.

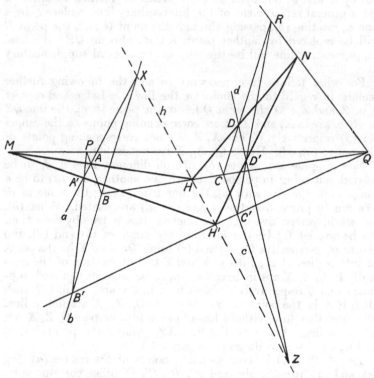

This proof is valid whether the lines *a, c* intersect or not, and also whether the lines *b, d* intersect or not. But it is necessary to consider whether the proposition is true when the lines *b, d, a* are not all different. The proposition has no meaning when *a* coincides with *b*; when *d* coincides either with *b* or with *a*, the line *h*, joining the intersection of *a, b* with the intersection of *c* and *d*, becomes the line *d*, and the previous construction fails. We must therefore consider the proposition independently. For this purpose it is necessary to define when two ranges on the same line are to be said to be related. If we take the natural course, and define two ranges on the same line as being related when they are both in perspective with the same range from different centres, we cannot at this stage give a proof to enable us to maintain the proposition above when the lines *a, d* are the same. We cannot in fact prove that if two ranges (*b*), (*d*) on different lines be related, then a range (*a*) on the line d which is in perspective with the range (b) is necessarily related to (d), in the sense that these two ranges (a), (d) are both in perspective with the same range. (The line upon which this last range lies would in fact meet the line d in a point belonging to both the ranges (a), (d); and until we have introduced the so-called complex points it is not necessarily true that the ranges (a), (d) on the line d have any common corresponding points.) Thus we shall define two ranges on the same line as being related when one of them, (a), is in perspective with a range, (b), which is related to the other, (d); it is then at once seen that a

range which is in perspective with the other is related to the one. Two ranges on the same line which are both in perspective with another range will then be related, though for not the most general possible reason.

This being understood, the above proposition is true when the lines a, d coincide, namely a range (a) on the line d is related to (d), when (a) is in perspective with a range (b) which is related to (d). And the proposition is also true when the lines b, d coincide, namely if two ranges (b), (d) on the same line are related, a range (a) in perspective with (b) is related to (d).

Thus finally the theorem that if we have a succession of ranges of which any consecutive two are related, then any range in perspective with the first is related to the last, remains true independently of the coincidence of the lines of any two related ranges.

Examples in regard to related ranges.

Ex. 1. If A, B, C, D be four points of a line, any two of the four ranges

$$A, B, C, D; \quad D, C, B, A; \quad C, D, A, B; \quad B, A, D, C,$$

which are obtained from the range A, B, C, D by an interchange of two of the points accompanied by an interchange of the other two points, are related.

Take any point O; join OA, OB, OC, OD; let any line through A meet the joins in A, B', C', D' respectively; let the line $D'C$ meet the join $OB'B$ in B''.

The two ranges A, B', C', D' and B, A, D, C are related, being both in perspective with the range B, B', O, B'', respectively from

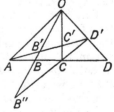

the centres C and D'; and the range A, B', C', D' is in perspective with A, B, C, D from the centre O. Thus (A, B, C, D) and (B, A, D, C) are related. From this the result follows.

Ex. 2. If on two non-intersecting lines there be two ranges A, B, C, D and A', B', C', D' which are related, so also are the ranges A, B, C, D and B', A', D', C', since (by Ex. 1) the latter is related to A', B', C', D'.

Thus if two ranges A, B, C, D and A', B', C', D', on two non-intersecting lines, be such that the four joins AA', BB', CC', DD' have a common transversal, then the four joins AB', BA', CD', DC' have a common transversal. It seems worth while to make a direct proof of this geometrical result, in verification of the foregoing theory.

The fact that the ranges A, B, C, D and A', B', C', D' are related,

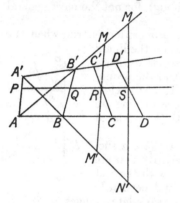

in accordance with what has been said above, is expressed by the fact that the lines AA', BB', CC', DD' have a common transversal, meeting them respectively, say, in P, Q, R, S. We wish to shew the existence of one or more common transversals of the cross joins AB', BA', CD', DC'. And in fact, if the transversal from R to AB' and $A'B$ meet these respectively in M and M', and the transversal from S, also to AB', $A'B$, meet these respectively in N and N', it can be shewn that the line MN', which is already a transversal of AB' and BA', is equally a transversal of CD' and DC'. The line $M'N$ is also a transversal of these four lines; and other transversals can be found by drawing from P, or from Q, the transversals to CD' and $C'D$. It can also be shewn that the point where the line MN' meets CD' lies on RA' and SA; and similarly the point $(MN', C'D)$ is the same as the point (RB, SB'), the point $(M'N, CD')$ the same as (RB', SB), and the point $(M'N, C'D)$ the same as (RA, SA').

For, first, the lines RA', SA are both in the plane $AA'Q$, and therefore meet, say in the point Y; while the line CD' lies in both the planes RCA' and SDA, and so meets both RA' and SA, but is not in the plane $AA'Q$. Thus CD' contains Y. Again, MN' lies in both the planes $RM'A'$ and SNA, and so meets both RA' and SA, but is not in the plane $AA'Q$. Thus MN' contains Y. Wherefore MN' meets CD' in Y.

Similarly RB and SB' meet, say in X, and $C'D$ lies in both the planes *BRC* and *SB'D'*, while *MN'* lies in both the planes *BRM'* and *SB'N*. Thus *MN'* meets *C'D* in *X*.

Also, *SB* and *RB'* meet, say in *T*, and *CD'* lies in both the planes *BSD* and *B'RC'*, while *M'N* lies in both the planes *BSN'* and *B'RM*. Thus *M'N* meets *CD'* in *T*.

Finally *RA* and *SA'* meet, say in *Z*, and *C'D* lies in both the planes *ARC* and *A'SD'*, while *M'N* lies in both the planes *ARM* and *A'SN*. Thus *M'N* meets *C'D* in *Z*.

The tetrad *A, B, B', A'* is in perspective with the tetrad *Z, X, T, Y*, from *R*, and in perspective with the tetrad *Y, T, X, Z*, from *S*. The four lines *MC, M'C', ND, N'D'* can be shewn to meet *BB'* in the same point, which is the harmonic conjugate of *Q* in regard to *B* and *B'*. Similarly the lines *MC', M'C, ND', N'D* all meet the line *AA'* in the harmonic conjugate of *P* in regard to *A* and *A'*.

Ex. 3. It is interesting to shew that, if four skew lines *AA', BB', CC', DD'* have three common transversals *ABCD, A'B'C'D', PQRS*, as in the preceding figure, we can, by the Propositions of Incidence alone, construct other transversals of these four skew lines, in infinite number, equally meeting every line having the three given transversals.

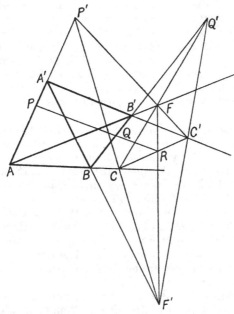

In the first place, if, on the line *AA'*, the point *P'* be the harmonic conjugate of *P* in regard to *A* and *A'*, and, on *BB'*, the point

Q' be the harmonic conjugate of Q in regard to B and B', while, similarly, on CC' and DD', the points R', S' are, respectively, the harmonic conjugates of R and S in regard to C, C' and D, D', then the points P', Q', R', S' are in line. And then the line $P'Q'R'S'$ is such a transversal of AA', BB', CC', DD' as was required.

For from R, through which passes the transversal CRC' of AB and $A'B'$, draw the transversal, FCF', of AB' and $A'B$, meeting these respectively in F and F'. Then the lines CF and $F'C'$ intersect one another, and, being respectively in the planes ABB' and $A'BB'$, they meet one another on the line of intersection, BB', of these planes, say at the point Q'. Similarly the lines $F'C$ and $C'F$ meet one another in a point, P', of the line AA'. It can be shewn that Q' and P' are the harmonic conjugates of Q and P respectively in regard to the couples B, B' and A, A'. For, from the construction made, the line $P'R$ meets $F'C'$ in the point harmonically conjugate to Q' in regard to F' and C'; and this point lies on the plane $P'RA'$, which is the plane $A'QA$. Wherefore, by perspective from A', the point Q is the harmonic conjugate of Q' in regard to B and B'. A similar argument is applicable to P and P'. Next, the line $P'Q'$, in the plane of FF' and CC', meets CC', and, by perspective from P', we see that it meets CC' in the point R' which is harmonically conjugate to R in regard to C and C'. If, now, from another point S, of the line PQR, there be drawn the transversal, DSD', to AB and $A'B'$, meeting these in D and D' respectively, and the transversal, GSG', to AB' and $A'B$, meeting these in G and G' respectively, we similarly shew that the lines DG, $D'G'$ meet in Q' and the lines DG', $D'G$ meet in P', and therefore that $P'A'$ meets DD' in the point S' which is the harmonic conjugate of S in regard to D and D'.

The line $P'Q'R'S'$ may be called appropriately the harmonic conjugate of the line $PQRS$ in regard to the lines $ABCD$ and $A'B'C'D'$. We may then similarly draw the harmonic conjugate of the line $A'B'C'D'$ in regard to the lines $ABCD$ and $PQRS$, and continue this process, obtaining from every three transversals of the four lines AA', BB', CC', DD' three others, and so on.

Another method of obtaining a fourth transversal of the four lines AA', BB', CC', DD', when we are given the three transversals $ABCD$, $A'B'C'D'$ and $PQRS$, may be explained, which also leads to an infinite number of transversals. The lines so constructed to meet the four given lines are not, however, shewn to meet any line with the same three transversals.

Let the lines $B'S$ and BR, which are both in the plane SBB', meet in the point X; also the lines $A'R$ and AS, which are both in the plane SAA', meet in the point Y. The line XP, in the former plane, SBB', will meet BB', say in Q'; the line YQ, in the plane SAA', will meet AA', say in P'. It can be shewn that the line $P'Q'$, which already meets AA' and BB', also meets CC' and DD'. In fact the four lines $P'Q'$, CC', XA, YB' meet in one point, as do the lines $P'Q'$, DD', YB, XA'. A simple method of verifying this, and similar results, arises later. For the present we may argue as follows.

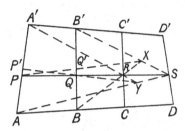

The lines XA, YB', in the plane ASB', intersect; the lines XA, $P'Q'$, in the plane XAP, intersect; the lines YB', $P'Q'$, in the plane YQB', also intersect. Hence the three lines $P'Q'$, XA, YB' lie in one plane or meet in one point; the former would involve that $B'Q'$ and AP', that is BB' and AA', were in one plane, and is excluded.

Again, the lines YB' and CC', in the plane $RC'A'$, intersect; and the lines AX, CC', in the plane BCR, intersect; while we have shewn that AX and YB' intersect. Hence the three lines CC', XA, YB' all lie in one plane, or meet in one point; the former would involve that $B'C'$ and AC, and therefore also BB' and CC', were in one plane, and is excluded.

Wherefore, as stated, the four lines $P'Q'$, CC', XA, YB' meet in one point.

A similar proof can be given for $P'Q'$, DD', XA', YB.

It can also be shewn that X lies on $C'D$, and that Y lies on CD'.

It cannot be proved at present that $P'Q'$ meets any other line meeting the three lines $ABCD$, $PQRS$, $A'B'C'D'$, beside the four AA', BB', CC', DD'; and it is in this respect different from the harmonic conjugate obtained above.

From $P'Q'R'S'$ we can find another line $P''Q''R''S''$ meeting AA', BB', CC', DD', just as we found $P'Q'R'S'$ from $PQRS$; and so on continually.

Ex. 4. Four planes, say α, β, γ, δ, which meet in a line, x, are said to form an axial pencil of planes. If these planes be met by any line, l, which does not meet the line x, respectively in the points A, B, C, D, and by any other line, l', not meeting x, respectively in the points A', B', C', D', then the ranges A, B, C, D and A', B', C', D' are related, since the joins AA', BB', CC', DD' have the common transversal x. We may thus speak of an axial pencil of planes, (ϖ), and a range of points, (m), as being related when the planes of (ϖ) are met by an arbitrary line, not meeting the axis of (ϖ), in a range which is related to (m).

With this definition, consider a tetrad of points A, B, C, D, not lying in a plane, and let any line, l, meet the four planes BCD,

CAD, ABD, ABC respectively in the points A_1, B_1, C_1, D_1. It can then be proved that the range A_1, B_1, C_1, D_1 is related to the pencil of planes lA, lB, lC, lD.

For let DB_1, in the plane CAD, meet AC in B'; let DC_1, in the

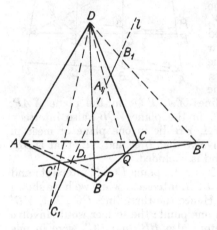

plane ABD, meet AB in C'; and let DA_1, in the plane BCD, meet BC in Q. Then the points C', D_1, Q, B' are in the plane Dl, and are thus in line. Let AD_1, in the plane ABC, meet BC in P.

Then the ranges G, D_1, A_1, B_1 and B, P, Q, C, lying on lines which do not intersect, are related, because the joining lines of corresponding points, BC_1, PD_1, QA_1, CB_1, all meet the line AD. Wherefore, by Ex. 2 above, the ranges C_1, D_1, A_1, B_1 and C, Q, P, B are related. The latter four points lie however, respectively, upon the planes lC, lD, lA, lB. This proves the result stated.

Ex. 5. We have shewn that Desargues' theorem follows by assuming the Propositions of Incidence for space of three dimensions. Conversely it can be shewn that, if Desargues' theorem for triads of points in one plane be assumed, there are aggregates of points and lines in this plane which combine together in accordance with laws corresponding to the Propositions of Incidence of space of three dimensions. Thus, also, beings of only two-dimensional experience might, in effect, be familiar with a three-dimensional geometry.

Take an arbitrary fixed point O of the plane. Any two points P, P' of the plane which are such that the line PP' passes through O, determine a point couple (P, P'), which, in the scheme we are explaining, plays the part of a point in a three-dimensional space of which the given plane is one plane. The couple (P', P) is to be distinguished from (P, P'). When P' coincides with P, the couple (P, P) represents the point P of the plane, regarded as belonging to the threefold space. When P' is at O, the couple (O, P), regarded as equivalent with (P, O), represents the point O itself, whatever P may be. Any two lines of the plane, taken in a definite order, give a line couple, which represents a line of the threefold space. Thus a line couple $(PQ, P'Q')$ is determined by two point couples (P, P') and (Q, Q'), and contains other point couples in infinite number, of which any two suffice for its determination. The condition that a line couple, (k, k'), should contain a point couple, (P, P'), is that the line k should contain the point P and also the line k' contain the point P'. Lastly, let the aggregate of a point couple (A, A'), and a line, l, of the given plane, represent a plane, any point couple (P, P') being regarded as lying on this plane when the lines PA and $P'A'$ meet on the line l. Thus if (Q, Q') be any other point couple of this plane, it follows, by Desargues'

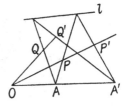

theorem, considering the two triads APQ, $A'P'Q'$, that the lines PQ, $P'Q'$ meet on the line l, so that the plane is equally represented by the point couple (P, P') and the line l; also any line couple $(PQ, P'Q')$ of the plane, determined by the two point couples (P, P') and (Q, Q'), is represented by two lines of the given plane which intersect on the line l. Any three point couples, (A, A'), (P, P'), (Q, Q'), determine the plane, which is given by the point couple (A, A') and a line l of the original plane containing the intersection of AP, $A'P'$, the intersection of AQ, $A'Q'$, and the intersection of PQ, $P'Q'$.

Any line couple, (k, k'), and plane (AA', l), have then a point couple in common. For, in the given

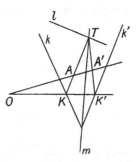

plane, with the lines k, k' and the points A, A', by drawing lines through O to meet k, k', we can, in virtue of Desargues' theorem, construct a line, m, passing through the intersection of k and k'. If this line meet the line l in T, the lines TA, TA', by their intersections with k, k' respectively, define a point couple (K, K'). This lies on the line couple (k, k') and also on the plane (AA', l), and is the point couple common to these.

Again, any two planes, (AA', k) and (BB', l), have a line couple in common. For take the line couple containing the two point couples (A, A') and (B, B'), which is represented by the two lines AB, $A'B'$ of the given plane; let these meet in M. Draw from M an arbitrary line in this plane, meeting the two given lines k, l, of this plane respectively in K and L; let KA, LB meet in Q, and KA', LB' meet in R. Then from Desargues' theorem, applied to the two triads A, Q, B and A', R, B', it follows that QR passes through O, so that (Q, R) is a point couple. As QA, RA' meet on k, and QB, RB' meet on l, this point couple belongs to both the planes. If another line ML_1K_1 be drawn through M, and Q_1 be the position thence obtained for Q, the application of Desargues' theorem to the triads L, K_1, Q_1 and L, K, Q shews that the line QQ_1 passes through the point H where k and l intersect. Similarly the point R describes a line passing through H. Thus the point couple (Q, R)

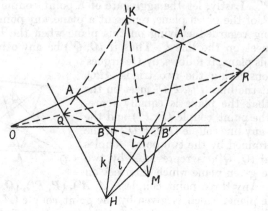

lies on a line lying in both planes, which is the line we desired to construct.

The theory may be continued. It may suffice to give one further example of its development, by drawing the line from a given point

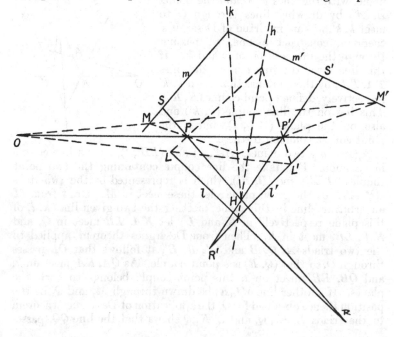

couple (P, P') to intersect each of two given lines represented by the line couples (l, l') and (m, m'). By drawing variable lines through O to meet the lines l, l' in L and L' respectively, and joining LP, $L'P'$, we define, by the locus of the intersection of these, a line h. Similarly lines through O meeting m and m' in M and M' respectively define a line k. Let these lines h, k intersect in H. Then let the lines HP, HP' meet l and l' respectively in R and R'; these will lie on a line through O and define a point couple (R, R'). So, let HP, HP' define by their intersections respectively with m and m' a point couple (S, S'). Then (HP, HP') are a line couple, representing a line, containing the given point couple (P, P'), which meets the line couple (l, l') in the point couple (R, R'), and meets the line couple (m, m') in the point couple (S, S'); as was desired.

Propositions of Incidence in space not limited to three dimensions. In general we assume that by means of two *points* a system of points is determined which can equally be determined by any two different points of the system. This system of points is said to lie on a *line*. A point is said *to be dependent upon*, or *to be in syzygy with*, two specified points, when it belongs to the system of points determined by these two. Later on we shall use the word *space* instead of the word *system*.

By means of three points, of which no one is dependent upon the others, a system of points is determined which can equally be determined by any three independent points of the system. The points of the line determined by any two points of the system all belong to the system. The system of points determined by three independent points is said to lie on a plane, or to be a twofold system. A point is said to be dependent upon, or to be in syzygy with, three specified independent points when it belongs to the system of points determined by these three.

By means of four independent points a system of points is determined which can equally be determined by any four independent points of the system. The points of a plane determined by any three independent points of the system, and therefore also the points of a line determined by two points of the system, all belong to the system. A point is said to be dependent upon, or in syzygy with, four independent points when it lies in the system determined by these four. Such a system is called a threefold system, or the points are said to lie on a threefold.

And so on in general, without limit.

Consider now an n-fold system which is determined by $(n + 1)$ points, these last being such as to contain three independent points, a fourth point independent of these, a fifth point independent of these four, and so on, and, finally, an $(n + 1)$th point independent of the preceding; in such an n-fold system there exist r-fold systems

($r < n$), each determined by $(r+1)$ independent points, of which, by hypothesis, every point belongs to the n-fold system. Two such contained systems, for the same or different values of r, may have points in common. Denote by S_r a system determined by $(r+1)$ independent points, and in particular by S_0 a single point. The general statement is that, *if r and h be two numbers both less than n but such that their sum is equal to or greater than n, then two systems, S_r and S_h, both contained in S_n, have always common a system S_{r+h-n}, but may have common a system S_k for which k is greater than $r+h-n$. When, as before, $r < n$ and $h < n$, but $r + h < n$, two systems S_r, S_h, both contained in S_n, have not always a system common. In general, with $r < n$, $h < n$, if two systems S_r, S_h, both contained in S_n, have common a system $S_{r+h-n+t}$, then there exists a system S_{n-t}, contained in S_n, which contains both S_r and S_h.*

We shew that these results are a logical consequence of one fundamental proposition, which, like the conception of a system of points determined by given points, we assume. This proposition is that, *if in the system S_n there be contained a system S_{n-1}, determined by n independent points of S_n, and also a system S_1, determined by two points of S_n which do not belong to S_{n-1}, then S_1 and S_{n-1} have one point in common,* so that every S_{n-1} contained in S_n has with every S_1 contained therein at least one common point. For instance we assume that, in a S_2 (or plane), any two S_1, (any two lines), have a point in common; and, in a S_3, any plane has common, with a line determined by two points not in this plane, a single point.

(1) From this proposition it follows that every S_r contained in S_n, where $r < n$, has a point in common with every S_{n-r} contained in S_n.

To prove this we remark that when $r = 2$ every S_{r-1} contained in S_n has a point in common with any S_{n-r+1} contained therein. Then we shew that, from the assumption that every S_{r-1} contained in S_n has a point in common with any S_{n-r+1} contained therein, it follows that every S_r contained therein has a point in common with any S_{n-r} so contained. Consider an S_r determined by the $r+1$ independent points $A_0, A_1, \ldots, A_r$, contained in S_n, and an S_{n-r} contained in S_n. It is supposed that S_r is not wholly contained in S_{n-r}; there is therefore at least one of the points $A_0, A_1, \ldots, A_r$ which is not contained in S_{n-r}; we may therefore suppose that the point A_r is not contained in S_{n-r}, and that the points $A_0, A_1, \ldots, A_{r-1}$ are independent points. The S_{n-r} is determined by $n - r + 1$ independent points, and we may suppose that A_r is independent of these, since $(n - r + 1) + 1$, or $n + 1 - (r - 1)$, is less than the greatest possible number, $n + 1$, of independent points in S_n (for $r = 2, 3, \ldots$). These $n - r + 1$ points together with A_r thus deter-

mine a Σ_{n-r+1} contained in S_n, which we may speak of as determined by S_{n-r} and A_r. On the other hand the points $A_0, A_1, ..., A_{r-1}$
determine a Σ_{r-1} contained in S_r. By what we have supposed, this
Σ_{r-1} and the Σ_{n-r+1} have a point in common (one at least), which
we may denote by P. This point P, being in Σ_{r-1}, is in S_r, as also
is A_r, and we can suppose P and A_r not to coincide; thus the
points P and A_r determine a Σ_1 lying in S_r. On the other hand
both P and A_r lie in the Σ_{n-r+1} determined by S_{n-r} and A_r, as
therefore also does the Σ_1. By the fundamental proposition, therefore, the Σ_1 and S_{n-r} have a point in common, say Q, which, as Σ_1
is contained in S_r, is a point common to S_r and S_{n-r}. This is the
point whose existence was to be shewn.

To explain the argument by a particular case suppose $n = 4$
and $r = 2$. We are to shew that any two planes in a space of four
dimensions have a point in common (one at least). Let these be
S_2, determined by the independent points A_0, A_1, A_2, and S_2', of
which S_2' does not contain A_2. Then A_2 and S_2' determine a Σ_3,
while A_0A_1 determine a Σ_1' contained
in S_2; this Σ_1' has with Σ_3 a point P
common. Both A_2 and P are in S_2, and
determine a Σ_1, the line A_2P. But A_2
and P are both in Σ_3, which contains
S_2'. Thus the line A_2P meets S_2', in Q,
which is then common to S_2 and S_2'.
Thus it appears as a consequence of the
facts that in an S_3 a line intersects a

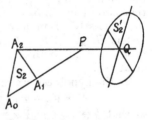

plane, and in an S_4 a line intersects an S_3, that, in an S_4, an S_2
intersects an S_2'.

(2) We next deduce, if $r < n$ and $h < n$, but $r + h \gtreqless n$, that an
S_r and an S_h have an S_{r+h-n} in common (at least). It is sufficient,
after (1), to prove this when $r + h > n$. For this we suppose that
an S_{r-1} and an S_h have an $S_{r+h-n-1}$ in common, which is true, by
(1), when $r = n - h + 1$.

We can, as in (1), suppose the S_r to be determined by a Σ_{r-1},
lying in S_r, and a point A, also lying therein. Then, because
$r \gtreqless n - h$, we can, in Σ_{r-1}, find a Σ_{n-h-1}; with A this will determine
a Σ_{n-h}, lying in S_r; by (1), this Σ_{n-h} will contain a point, P, of
S_h; this point P lies then in S_r and S_h. By hypothesis the Σ_{r-1}
and S_h have a $\Sigma_{r+h-n-1}$ in common; this, lying in Σ_{r-1}, also lies in
S_r. Assuming, for a moment, that P does not lie in the $\Sigma_{r+h-n-1}$,
this point P and the $\Sigma_{r+h-n-1}$, together, determine an S_{r+h-n}, and
this lies both in S_r and S_h because both P and $\Sigma_{r+h-n-1}$ do so. This
establishes the statement made.

That the common point, P, of S_h and the Σ_{n-h} determined by
Σ_{n-h-1} and A, does not necessarily lie in the $\Sigma_{r+h-n-1}$ common to

Σ_{r-1} and S_h seems clear; for we can suppose P determined by S_h and A and Σ_{n-h-1}, before we specify the $r+h-n$ points which, taken with Σ_{n-h-1}, determine the Σ_{r-1}; if P, which is common to S_h and Σ_{n-h}, were always in Σ_{r-1}, it would be in Σ_{n-h-1}; this however has, in general, no point common with S_h.

For an illustrative example, let $n=3$, $r=2$, $h=2$; the theorem will be that in an S_3, a space of three dimensions, an S_2 has with an S_2' a line common. We suppose the S_2 determined by a line Σ_1 and a point A. In this line Σ_1 take a point Σ_0; this point, with A, determines a line Σ_{n-h}, lying in S_2. This line, Σ_{n-h}, contains a point P of S_2', lying in S_2. The line Σ_1 has, with S_2', a common point Σ_0', also in S_2. As P and Σ_0' are different, they determine a line, $P\Sigma_0'$, lying both in S_2 and S_2'. Thus the hypothesis that a line and an S_2', lying in an S_3, have a point in common, involves that two planes in S_3 have a line in common.

(3) Lastly, if in a system S_n, an S_r and S_h have common a system $S_{r+h-n+t}$, then there exists a system S_{n-t}, contained in S_n, which contains both S_r and S_h. It is not assumed that $r+h$ is as great as n.

By the hypothesis we must have

$$r+h-n+t \gtreqqless r, \text{ and } r+h-n+t \gtreqqless h,$$

so that $h \lesseqqgtr n-t$ and $r \lesseqqgtr n-t$. We can take, in $S_{r+h-n+t}$, independent points in number $r+h-n+t+1$; these, with $n-t-h$ points, independent of the former, taken in S_r, will furnish $r+1$ independent points by which S_r may be determined; the same $r+h-n+t+1$ points in $S_{r+h-n+t}$, with $n-t-r$ points independent of those already taken in S_h, will furnish $h+1$ points by which S_h may be determined. The total number of independent points thus employed is, however, only

$$(r+h-n+t+1)+(n-t-h)+(n-t-r)$$

or $n-t+1$. The system determined by these, which is a S_{n-t}, contains then both S_r and S_h.

For example, for $n=4$, $r=h=2$, two planes in a fourfold space have not a line in common unless they both lie in the same space of three dimensions.

Examples of the Propositions of Incidence in space of four and five dimensions.

Ex. 1. In space of four dimensions, two lines which have no point in common, being each determined by two points, determine together a space of three dimensions, defined by four points two on each line. This threefold space has in common with any arbitrary third line of the fourfold space a single point; from this point can be drawn, in this threefold space, a single transversal line of the two former lines. Thus we can shew that three general lines in space of four dimensions are all met by one other definite line. If the three lines lie in a threefold space they have, of course, an infinite number of common transversals.

Ex. 2. In general a line and plane in a fourfold space have no point in common. If they have a point *O* in common, this with another point *P* of the line, and two other points *A*, *B* of the plane not lying in line with *O*, make four points, which determine a threefold space containing both the given line and plane.

But through one of three lines in fourfold space, having one common transversal, can be drawn a single plane to meet each of the other two lines in a point, that namely containing the chosen line and the common transversal line of the three given lines.

It follows that, in space of four dimensions, ∞^1 planes can be drawn through an arbitrary point *O* to meet each of three given lines of general position. For a plane can be drawn containing the line joining *O* to an arbitrary point of one of the three given lines which shall meet each of the other two. We can further specify three other lines each of which is met by every one of these planes. For consider one of these planes, meeting the three given lines *a*, *b*, *c*, respectively in *A*, *B*, *C*. In this plane the lines *AB* and *OC* intersect one another, say in the point *P*. And this point *P* is clearly on the line which is the intersection of the threefold defined by the lines *a*, *b* and the plane *Oc*. There are two other lines which arise by replacing *c*, in this description, by *a* or by *b*.

Ex. 3. It has already been remarked above that the necessary condition for two planes in fourfold space to have a line in common is that these planes should lie in a threefold space. This space will be determined by two points of the line and two further points, one on each plane.

Ex. 4. Three lines in most general position, being determined by six points, two on each line, lie in at most five dimensions; if the lines have a single common transversal line, they lie in the fourfold space determined by two points of this and a further point on each of the lines; if they have two common non-intersecting transversal lines they lie in the threefold space determined by these. Similarly a line and a plane, determined by two points of the line and three points of the plane, lie in a fourfold space, unless the line and plane have a common point. Further, two planes are determined by three points on each, and lie in at most five dimensions; if the planes have one common point they lie in four dimensions; if they have two common points they lie in three dimensions.

Ex. 5. It was seen that in three dimensions a definite line can be drawn from a point to meet two general non-intersecting lines. A corresponding result is that, in five dimensions, a definite plane can be drawn from a point to meet three general lines. For let the lines be given respectively by the pairs of points A, A'; B, B'; C, C'. We may suppose these to be six independent points determining the fivefold space. Let O be another point of this space. By the five points O, A, A', B, B' a fourfold space is determined; by the five points O, A, A', C, C' another such. The common points of these two spaces determine a threefold space, S_3, of which O is one point and AA' is one line. In the fivefold space this threefold space, S_3, meets the fourfold space determined by the points O, B, B', C, C' in a plane, ϖ; this plane, ϖ, contains O, and, being in S_3, it meets the line AA' in a point. It is however the common plane of the three fourfold spaces O, A, A', B, B'; O, A, A', C, C'; O, B, B', C, C', and thus by symmetry equally meets the lines BB' and CC'. It is therefore such a plane as is desired. And it is the only such; for any plane through O meeting the lines AA' and BB' contains three points of the fourfold space O, A, A', B, B' and therefore lies in this space.

In consequence there are ∞^1 planes meeting four lines of general position in space of five dimensions. Denoting the lines respectively by a, b, c, d, and the points in which such a plane meets these respectively by P, Q, R, S, the point of intersection of the lines QR and PS is in the threefold space determined by the lines b, c, and also in the threefold space determined by the lines a, d. It is therefore on the line which is the intersection of these threefold spaces $\{b, c\}$ and $\{a, d\}$. The plane in question, meeting the four lines a, b, c, d, therefore meets this line; as well, similarly, as the line of intersection of the threefold spaces $\{c, a\}$ and $\{b, d\}$, and the line of intersection of the spaces $\{a, b\}$, $\{c, d\}$.

Ex. 6. The first part of Desargues' theorem was concerned with two triads of points, A, B, C and A', B', C', in threefold space, with the property that the joins of corresponding points, namely the lines AA', BB', CC', had a point, O, in common. Consider now, correspondingly, in fourfold space, two tetrads A, B, C, D and A', B', C', D', likewise in perspective by lines from a point, O, it being supposed that A, B, C, D are not in a plane and, therefore, determine a threefold space, while A', B', C', D' similarly determine another such space, these two spaces intersecting in a plane, say ϖ. It then follows, as in Desargues' theorem, that corresponding joining lines of points in these tetrads, as for example the lines BC and $B'C'$, meet in a point lying in the plane ϖ. Thence, two corresponding planes of the two tetrads meet in a line of the plane ϖ; for instance the planes A, B, C and A', B', C' have common the three points $(BC, B'C')$, $(CA, C'A')$, $(AB, A'B')$, which, as the planes ABC, $A'B'C'$ are distinct, must be in line. The six points of intersection of corresponding lines are then on four lines of the plane ϖ.

Ex. 7. The results of Examples 2 and 5 may be generalised — and the conclusions will be of interest for an application to the theory of related ranges of points.

We prove that in space of an odd number of dimensions, S_{2n-1}, a definite S_{n-1} can be drawn through a point, O, to meet n lines of general position, each in one point. And hence, that there are ∞^1 spaces S_{n-1} meeting $(n+1)$ lines of general position; and further, when $n > 2$, that each of these S_{n-1} meets the line which is the intersection of the S_3, containing any two of the $(n+1)$ given lines, with the S_{2n-3} containing the remaining $(n-1)$ of the given lines. There are $\frac{1}{2}n(n+1)$ lines of this description.

We prove, also, that in space of an even number of dimensions, S_{2n}, a definite S_{n-1} can be drawn to meet $(n+1)$ lines of general position. And hence, that there are ∞^1 spaces S_n passing through a point, O, which meet $(n+1)$ lines of general position; and further, when $n > 1$, that each of these S_n meets the line which is the intersection, of the S_3 containing two of the $(n+1)$ given lines, with the S_{2n-1} containing O and the remaining $(n-1)$ of the given lines. There are $\frac{1}{2}n(n+1)$ lines of this description.

Consider a space S_{2n-1}, and therein a point, O, and n lines of general position. Let a be one of the lines; the remaining $(n-1)$ lines, each determined by two of its points, together with O, depend upon $(2n-1)$ points, which, as we provisionally assume, may be regarded as independent. These determine then a S_{2n-2} containing O and these $(n-1)$ lines, which we speak of as determined by O and these lines. In the S_{2n-1}, this S_{2n-2} meets the line a in a point, say A. There are other $(n-1)$ spaces, also of dimension $(2n-2)$, each of which contains the point O, the line a, and all of the remaining given lines except one. The n spaces, S_{2n-2}, thus defined, meet in a S_{n-1}, which contains the point A, this being on every one of the spaces. By symmetry this S_{n-1} contains a point of each of the n given lines; and it passes through O. Conversely a S_{n-1} through O, which meets each of the n given lines, has n points in common with a S_{2n-2} containing O and any $(n-1)$ of these lines; such S_{n-1} thus coincides with that found, which is therefore unique.

It follows that there are ∞^1 spaces S_{n-1} meeting $(n+1)$ lines of general position in S_{2n-1}, namely one such through every point of a selected one of the $(n+1)$ given lines. Let one such S_{n-1} meet two of these given lines, say a_1 and a_2, respectively in A_1 and A_2. The $(n-1)$ intersections of this S_{n-1} with the other $(n-1)$ given lines determine a S_{n-2} lying in the chosen S_{n-1}; and the line A_1A_2, lying in the S_{n-1}, meets this S_{n-2} in a point, say P. This is a point of the S_{n-1}. The line A_1A_2, however, lies in the S_3 determined by the lines a_1 and a_2; and the S_{n-2} contains $(n-1)$ points of the S_{2n-3} containing the other $(n-1)$ given lines, and so lies in this S_{2n-3}. Thus the point P is in the space common to the S_3 and S_{2n-3}; this common space is of dimension $(2n-1)-(2n-4+2)$ or 1, namely it is a line. This proves the result stated for a space S_{2n-1}. The method of proof is provisional, since we have, for simplicity, assumed without examination various sets to consist of independent points. The same remark applies to the case of a space S_{2n}, which we now consider.

In a space S_{2n} suppose $(n+1)$ lines of general position to be given. Let a be one of these lines. The remaining n lines determine a S_{2n-1}, which will meet the line a in a point, say A. The space common to this S_{2n-1} and to the other n spaces of dimension $2n-1$, each determined by the line a and $(n-1)$ of the remaining lines, is of dimension $2n-(n+1), =n-1$. This S_{n-1} contains the point A, which is on each of the $(n+1)$ spaces S_{2n-1}; and thus, by symmetry, this S_{n-1} meets every one of the $(n+1)$ given lines. Conversely, any existing S_{n-1} meeting these lines, as it contains n points of each S_{2n-1} containing n of these lines, lies in all the $(n+1)$ spaces S_{2n-1}; the S_{n-1} found is therefore unique. Hence we deduce that a single space S_n exists containing an arbitrary line, b, which meets n other lines of general position. For this line, b, and these n others, are all met by a certain S_{n-1}, as we have shewn; this S_{n-1} is determined by n points which, with a point of the line b not lying on the S_{n-1}, give $(n+1)$ points determining a S_n containing b and the S_{n-1}; this S_n therefore meets these n other lines each in a point.

It follows that ∞^1 spaces S_n can be drawn through an arbitrary point O to meet $(n+1)$ lines of general position. Take one of these S_n; let a_1, a_2 be two of the $(n+1)$ lines, the aggregate of the remaining $(n-1)$ lines being denoted by (c). On these lines (c) there are the $(n-1)$ points in which they are met by this S_n, and this also contains the point O. These n points determine a S_{n-1} lying in this S_n, and this is intersected in a point by the line, say t, also lying in this S_n, joining the two points where this S_n meets a_1 and a_2. Thus we have a point, say P, of S_n not on any of the $(n+1)$ given lines. The S_{n-1} contains n points of the S_{2n-2} determined by the $(n-1)$ lines (c) and the point O, and the line t lies in the S_3 determined by the two lines a_1, a_2. The point P therefore lies on the intersection of this S_{2n-2} and S_3, and these, in S_{2n}, meet in a space of $2n-(2+2n-3)$ dimensions, that is in a line. The ∞^1 spaces S_n thus meet the line which is the intersection of the S_3, containing two of the $(n+1)$ given lines, and the S_{2n-2} determined by O and the $(n-1)$ remaining lines. There are $\frac{1}{2}n(n+1)$ lines of this description.

Ex. 8. We have said that two ranges of points on two lines in threefold space are related when they are both in perspective with the same range on another, intermediary, line, from two appropriate centres. The threefold space may be determined by the two centres of perspective and two points of the intermediary line. As a particular case the two ranges may lie on lines which are in a plane. When this is not so the ranges are related if the joining lines of corresponding points of the ranges all meet another line. The definition gives a geometrical construction by which we may pass from a point of either range to the corresponding point of the other range *depending only on the Propositions of Incidence.* Suppose that, in a space which is of higher than three dimensions, we have a set of lines arranged in order, say $a, b, c, \ldots, k$, each containing a range of points, and that the ranges (a), (b) are related by the preceding construction, which will be confined to the threefold (or twofold) space determined by the lines a, b, also the ranges (b), (c) related by this construction, in the threefold space (b, c), and so on for every consecutive two. Then we may enquire whether the ranges (a), (k) are related by the preceding construction. Next suppose that in a space of more than three dimensions we have two ranges (a), (k) for which we have a geometrical construction, only depending on the Propositions of Incidence, defining uniquely the point of either range from a corresponding point of the other. We may then enquire whether two such ranges are related according to the preceding definition. If the answer to this query be affirmative, that to the former will also be.

We do not enter into this enquiry now. But the preceding example gives a case which is illustrative. In four dimensions a geometrical correspondence is effected between the points, P, Q, of two lines, a, b, by means of planes containing these points drawn through a point O to meet a third line, c. We have shewn that then the joining line PQ meets the plane Oc in the points of a line lying in the threefold determined by a and b. In five dimensions a geometrical correspondence is effected between the points, P, Q, of two lines, a, b, by means of planes containing these points which meet two other lines, c, d. We have shewn that then the joining line PQ meets the threefold space determined by c, d in the points of a line lying in the threefold space determined by a and b. And we have shewn that this can be generalised to space of higher dimension.

Ex. 9. Another example interesting to consider in connexion with the enquiry suggested in the last example arises for the planes which meet four lines of general position in space of four dimen-

sions. One such plane can be drawn passing through two arbitrary
points one on each of two of the four given lines; so that we may
say there are ∞^2 such planes. It will be proved subsequently, with
the help of an assumption in addition to the Propositions of Incidence, that all these planes meet another line.

Let the given lines be denoted by a, b, c, d, and the line which
meets a, b, c be denoted by d', the lines meeting the other sets of
three of the four given lines being denoted respectively by a', b', c';
then it can be shewn from the Propositions of Incidence only that
the four threefolds determined respectively by the pairs of lines
(a, a'), (b, b'), (c, c') and (d, d') do not meet in a point merely, but
in a line. It is this line which is shewn to be met by any plane
meeting the lines a, b, c, d.

On the Principle of Duality in general. We have remarked
that in three dimensions the Propositions of Incidence establish a
duality whereby, to every geometrical result involving points, lines
and planes, there corresponds a result involving respectively planes,
lines and points. A similar duality arises from the Propositions of
Incidence in space of any number of dimensions. If the space be
of n dimensions the correspondence between spaces contained therein
is between a space of dimension r and a space of dimension $n - r - 1$.
Thus to a point corresponds a S_{n-1}; to the line determined by two
points corresponds the S_{n-2} which is the intersection of two spaces
S_{n-1}, and so on. To our fundamental proposition that a line meets
a S_{n-1} in a point corresponds then the statement that a S_{n-2} and a
point together determine a S_{n-1}, which was a more fundamental
part of our description of what is a space, or system.

Thus, in four dimensions, to a point corresponds a threefold and
to a line corresponds a plane; and conversely in each case. For
example, as two planes in four dimensions have in general a point
in common, so two lines determine a threefold; the fact that when
the planes have a line in common they lie in a threefold, becomes
the fact that when two lines lie in a plane they have a point in
common. The fact that a line and a plane which meet in a point
lie in a threefold is evidently dual to its converse. Again as there
is a line meeting each of three general lines in a point, so there is
a plane meeting each of three given planes in a line; the fact that
three lines lying in a threefold are met, each in a point, by an
infinite number of lines lying in this threefold, becomes the fact
that if three planes have a point in common an infinite number of
planes can be drawn through this point each meeting each of the
three given planes in a line. The result we have discussed above,
that there is an infinite number of planes through a given point
meeting each of three given lines, of general position, in a point,
becomes the statement that of the lines meeting three given planes

of general position there is an infinite number lying in an assigned threefold; these are evidently the common transversals of the lines in which the three given planes meet the threefold.

In five dimensions a line is dual to a threefold and a plane is dual to a plane. For example, the fact that two lines in a plane have a point in common is dual to the fact that if two threefolds, instead of meeting in a line, meet in a plane, then they lie in a four-fold space.

On projection from one space to another. It is frequently possible to make a correspondence by geometrical construction between the points of one space, α, and the points of another space, β, a single arbitrary point of either space being thus related uniquely to a single point of the other space. When through certain fixed points, or spaces, γ, variable spaces, ϖ, can be drawn, each of which intersects the space α in one point, and is made unique by the assignment of this point, and then intersects the space β in one point, and if the same is true when instead of α, β we say respectively β, α, then such a correspondence may be called a projection. The process will have so many applications that it seems unnecessary to do more at this stage than to refer to it as one of the fundamental things.

SECTION II. THE CORRESPONDENCE OF THE POINTS OF TWO LINES. PAPPUS' THEOREM

Preliminary remarks. In using the words point, line, plane, for the entities of our theory, it is clearly intended that the properties obtained shall not be widely different from those ordinarily supposed to belong to the entities, given or suggested by experience, to which these familiar names are attached. In particular, the fullness, or multiplicity, of the points that are possible on a line must not be widely different in the two cases. As will appear quite adequately in the sequel, this condition is not secured by the Propositions of Incidence alone; these leave us free to regard as distinct, because differently obtained, points which, from the suggestions of our experience, we should regard as coincident. This experience takes account in fact of other ideas than those we have embodied in the Propositions of Incidence, in particular of the idea of one point of the line being *between* two others. This idea is considered in Chapter II. As has been stated in the Introductory Remarks to this volume, we do not finally adopt this idea as fundamental; the present Section is intended to lead up to the adoption of another assumption which will enable us to avoid this.

The question of the identification of points upon a line we merge in the question of the identity of the points of two ranges of points on the line; and this we regard as a particular case of the question when two ranges of points on two different lines can be regarded as being in unambiguous correspondence, so that the specification of a point of either line leads to a definite point of the other. One way in which such a correspondence arises is suggested by our preceding work; the points of two ranges which are in perspective from a point are evidently in such unambiguous correspondence. It is natural then to make the condition that a method of correspondence adopted shall be such that ranges in perspective are a particular case; and hence that ranges which are in correspondence by means of a succession of any number of perspectivities shall also be in correspondence in virtue of the definition which we adopt. In a phraseology we have used, and by a result obtained in Section I, this comes to saying that two ranges which are *related*, in particular in perspective, shall correspond with one another. A particular consequence of this condition is that if four points of one range be in the harmonic relation to one another, this shall be true also of the four corresponding points of the other range. But, of four points in this harmonic relation, three can be taken arbitrarily. Thus the method of correspondence which we adopt must be of such generality that, having taken any three points upon one of two lines, a correspondence is possible in which any three arbitrarily chosen points of the other line correspond respectively to these; and the same must be true of two ranges upon the same line which correspond to one another. Now, we can find upon any line, b, a range which is related, in the sense used in Section I, to a given range upon another line, a, so that three arbitrary points of b shall correspond to three specified points of the given range (a); if the Propositions of Incidence were sufficient to secure an unique point of b to correspond then to an arbitrary point of the given range (a), these Propositions might be a sufficient basis for such a theory of correspondence. Unfortunately this is not so. On the other hand a range on one line, related to a given range on another line, cannot be constructed so that four arbitrary points of the constructed range shall correspond to four given arbitrary points of the original range.

Hence either we must give up the suggestion that the method of correspondence adopted, ultimately with a view to the identification of the points of a line, shall lead us to regard related ranges as being in correspondence; or else a statement that two ranges are in correspondence because they are related must always be accompanied by a statement of a particular construction rendering the method of relation unambiguous; or else finally we must add to the Propositions of Incidence some further general Proposition which, probably by acting as a restriction of what points we regard as possible on a line, shall secure that the correspondence of the points of two related ranges is unique when three points of the one are given as corresponding to three points of the other. It is the last alternative which we adopt. As will be seen, a geometrical theory can thus be framed which, avoiding many particular assumptions, obtains results exactly corresponding to those usually based on these assumptions. The merit of pointing out, in part, the very remarkable fact that such a course should be possible belongs originally to H. Wiener, *Jahresbericht d. Deutschen Math.-Verein*, Erster Bd., 1890–91, p. 47. His statement relates to geometry in a plane. For subsequent development cf. F. Schur, *Math. Annal.* LI, 1899, p. 402 and Hilbert, *Gauss-Weber Festschrift, Grundlagen der Geometrie*, 1899.

Conditions that the correspondence of two related ranges should be of freedom three. Consider, in space of three dimensions until the contrary is said, two lines, a, b, which do not intersect. Upon the former take three arbitrary points, A, A', A'', and upon the latter three arbitrary points, B, B', B''. Join AB; upon this take any point C. From C draw the transversal, $CC'C''$, to meet the join $A'B'$ in the point C', and the join $A''B''$ in the point C''; denote this transversal by c. Then, if A''' be any fourth point on the line a, the transversal, $A'''B'''C'''$, drawn from A''', to meet the line b in B''' and the line c in C''', makes correspond to the arbitrary point A''' of a the definite point B''' of b. Namely the assignment of three points, B, B', B'', of b, to correspond respectively to three given points, A, A', A'', of a, *with the additional assignment of a definite point C of the line AB*, makes correspond to any range $(AA'A''A''' ...)$ of the line a, a definite range $(BB'B''B''' ...)$ of the line b.

But it does not follow that this correspondence is independent of the position of C upon the line AB, when A, A', A'' and B, B', B'' remain the same. Let D be any other point of the line AB, and draw the transversal d to meet the line $A'B'$ in the point D', and $A''B''$ in D''; then denote by B_3 the point in which the transversal $A'''B_3D_3$, drawn from A''' to meet the lines b and d, meets the line b. It does not follow from the Propositions of Incidence alone that the point B_3 coincides with B'''.

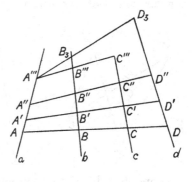

If B_3 do coincide with B''', then the line $A'''B'''C'''$ coincides with the line $A'''B_3D_3$ and meets all of a, b, c, d. If such coincidence holds for all positions of A''' upon a, the four lines a, b, c, d, simply in virtue of having three transversals, $ABCD, A'B'C'D', A''B''C''D''$, will have an infinite number of transversals, namely, what is important, one passing through every point of the line a. And, if it be true generally that any four lines, of which no two intersect, (lying in a threefold space), which have three transversals meeting all of them, have also a common transversal passing through any point of one of them, then, in the figure just considered, the four lines $ABC, A'B'C', A''B''C'', A'''B'''C'''$ will be intersected by the transversal drawn to the first three of them from *any* point, D, of the first.

We thus have two equivalent conditions which are necessary, and are also sufficient, in order to secure that the correspondence of two related ranges, on two non-intersecting lines a, b, may be uniquely determined by the assignment of three points of one of these lines to correspond respectively to three assigned points of the other, namely:

1. That if four skew lines have three common transversals, they should possess a common transversal through any point of either one of the four lines.

2. That if three lines, a, b, c, of which no two intersect, be each met by three other lines, $ABC, A'B'C', A''B''C''$, then every transversal of the first three lines, a, b, c, should meet every transversal of the second set of three lines.

Brianchon hexagons and another form of the conditions. This last can be replaced by another equivalent condition. Consider six points, not lying in a plane, joined in a definite order, P, Q', R, P', Q, R', to form what we describe as a skew hexagon. It is easy to see that, if the three lines PP', QQ', RR', joining opposite angular points of the hexagon, meet in a point, say O, then any

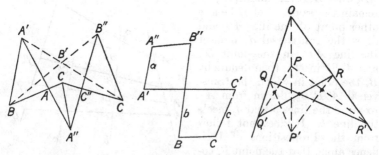

two opposite sides of the hexagon, such as $PQ', P'Q$, or $Q'R, QR'$, or $RP', R'P$, intersect one another. And conversely that, if every pair of opposite sides is in a plane, the three diagonals meet in a point. A hexagon with this particularity we may distinguish by calling it a *Brianchon* hexagon. Such a hexagon is constituted by six points of the figure we were discussing, namely, in order, A', A'', B'', B, C, C'. The three pairs of opposite sides of this hexagon, namely $(A'A'', BC)$, $(A''B'', CC')$, $(B''B, C'A')$, meet, respectively, in the points A, C'', B'. The condition (2) above can then be stated in the form

3. That any transversal of one set of alternate sides (here the lines a, b, c) of a Brianchon hexagon should intersect every transversal of the other set of alternate sides (here the lines BC, $C'A'$, $A''B''$).

Pappus' Theorem. Is true if the conditions are satisfied. These three conditions are, in virtue of the Propositions of Incidence, equivalent to one another. We proceed to shew that they are also equivalent to a certain proposition of geometry in a plane; that is, that this proposition follows if any one of the three conditions is supposed always to be satisfied, while, conversely, the truth of this proposition would secure that every one of these conditions is satisfied. The proposition in question is

4. If L, M, N be any three points of a line, and L', M', N' be any three points of another line, intersecting the former, then the three points of intersection of cross joins

$$(MN', M'N), \quad (NL', N'L), \quad (LM', L'M)$$

are in line.

This proposition is stated in the συναγωγή of Pappus (? 340 A.D. See Ch. Taylor, *Ancient and Modern Geometry of Conics*, Cambridge, 1881, p. liii, and the works of T. L. Heath on the geometry of the Greeks referred to in the Bibliography at the end of this volume). We shall refer to it as Pappus' Theorem.

We shew, first, that Pappus' theorem follows if the condition (2) above be satisfied.

Let O be the point of insection of the lines LMN and $L'M'N'$, which we may denote respectively by e and e'. Through the points L and L' draw two lines, respectively a and a', intersecting in a point P, not lying in the plane of the lines e and e', but otherwise arbitrary. Through the point M draw any line, b, intersecting the line a', say in the point C. This line b will not intersect the line a. or the line $L'M'N'$. since.

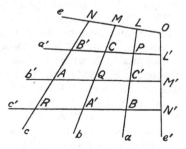

otherwise, a' would intersect the line LMN and be in the plane of
e and e'. Then through the point M' draw the transversal to the
lines a and b, say the line b', meeting the lines a, b respectively in
C' and Q. This line b' will not meet the line a', or the line LMN,
since, otherwise, the line a would intersect the line $L'M'N'$ and be
in the plane of e and e'. Lastly, draw from the point N the transversal to the lines a' and b', say the line c, meeting a' and b', respectively, in the points B' and A. This line c will not meet any
one of the lines b, a, $L'M'N'$, since, otherwise, the line a' would
meet the line LMN. And draw from the point N' the transversal
to the lines a, b, say the line c', meeting these, respectively, in B
and A'. This line c' will not meet any one of the lines b', a', LMN,
since, otherwise, the line a would meet $L'M'N'$. The fact that no
two of the four lines a', b', c', LMN intersect, and no two of the
four lines a, b, c, $L'M'N'$ intersect, is thus a consequence of taking
P to be any point not lying in the plane of the two lines e and e'.

If now we assume the condition (2) above to be satisfied, it follows
that the two lines c, c' intersect one another, say in the point R.
Let this be so.

The line QR does not meet either the line e or the line e'; since,
in the former case, b and c would intersect, and, in the latter case,
b' and c' would intersect; thus QR does not lie in the plane $[e, e']$,
which contains the two lines MN' and $M'N$. But QR intersects the
line MN', these two lines, QR, MN', being both in the plane $[b, c']$;
and QR intersects $M'N$, these two being both in the plane $[b', c]$.
Therefore the line QR must pass through the point of intersection
of the lines MN' and $M'N$; this intersection may thus be defined
as the point of intersection of the line QR with the plane $[e, e']$.
The points $(NL', N'L)$ and $(LM', L'M)$ are, similarly, the intersections, respectively, of the lines RP and PQ with the plane $[e, e']$.
The points P, Q, R are not in line, since the points L, M, N,
L', M', N' are supposed to be of such general positions that the
three points $(MN', M'N)$, $(NL', N'L)$, $(LM', L'M)$ are different.
These three points are therefore in line, namely in the line in which
the plane PQR meets the plane $[e, e']$.

Thus Pappus' theorem is shewn to follow if the condition (2) is
satisfied, and therefore follows equally if either of the equivalent
conditions, (1) or (2), is satisfied.

**Pappus' Theorem. The conditions follow if Pappus'
Theorem be assumed.** Conversely we now shew that if Pappus'
theorem be assumed, then condition (2) is satisfied.

Let a, b, c be any three lines, of which no two intersect, and
a', b', c' any other three lines of which, equally, no two intersect;
let every one of the lines a', b', c' intersect every one of a, b, c, the
respective intersections being P, C, B' on a', and C', Q, A on b',

and B, A', R on c'. Draw any further transversal of the lines a, b, c, say the line e, meeting these respectively in L, M, N, this line not meeting the lines a', b', or c'. Then, from any point, N', of the line c', draw the transversal, say e_1, of the lines e and a', meeting these in O and L_1, respectively. As c' does not meet e, this line e_1 will not meet a, or b, or c. As before, the line RP does not lie in the plane $[e, e_1]$, but

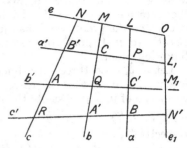

meets both the lines $N'L$ and NL_1, lying with these, respectively, in the planes $[a, c']$ and $[a', c]$; so that the lines $N'L$ and NL_1 meet in the point where RP meets the plane $[e, e_1]$. The transversal e_1 does not lie in the plane $[a, b']$, since it does not meet the line a; let it meet this plane in the point M_1. The line PQ does not lie in the plane $[e, e_1]$, but meets the lines LM_1 and L_1M, lying with these, respectively, in the planes $[a, b']$ and $[a', b]$. Thus LM_1 and L_1M meet in the point in which the line PQ meets the plane $[e, e_1]$. The points P, Q, R do not lie in line, since the lines $N'L$, NL_1, L_1M do not meet in a point, unless N' be taken at B. There is thus a line in which the plane PQR meets the plane $[e, e_1]$. The application of Pappus' theorem to the two triads L, M, N and L_1, M_1, N' would lead to a line in the plane $[e, e_1]$ containing the three points

$$(LM_1, L_1M), \quad (MN', M_1N), \quad (NL_1, N'L);$$

of these the first and third have been shewn to lie, respectively, on PQ and PR; thus the second, the point (MN', M_1N), lies in the plane PQR. The line MN' does not lie in this plane; for, if so, it would intersect PQ, in a point other than Q, and, then, PQ, having two points in the plane $[b, c']$, would lie in this plane, so that PB, that is the line a, would intersect the line b. But MN', in the plane $[b, c']$, meets QR; therefore, as the point (MN', M_1N) lies in the plane PQR, it follows that M_1N intersects QR. Hence M_1 lies in the plane NQR, or $[b', c]$. And M_1 was constructed to be in the plane $[a, b']$. Thus, M_1 is on the line b', and the transversal e_1, drawn from N' to meet e and a', is the same as the transversal which can be drawn from N' to meet a' and b'; this last transversal therefore meets e. And this is the same as that the condition (2) is satisfied.

Alternative proof of the equivalence of Pappus' theorem with the conditions. The six points $AB'CA'BC'$ constitute a Brianchon hexagon, the pairs of opposite sides

$$(AB', A'B), \quad (B'C, BC'), \quad (CA', C'A)$$

being intersecting lines, meeting, respectively, in R, P, Q. The joins of opposite angular points, namely the lines AA', BB', CC', therefore meet in a point. Taking account of this, we may substitute for the proof which has just been given, that condition (2) is satisfied if Pappus' theorem be assumed, and conversely, the following proof.

Let a, b, c be three skew lines, that is, lines of which no two intersect, and a', b', c' be three other skew lines, all intersecting every one of a, b, c. Denote the nine points

$$(a, a'), \ (a, b'), \ (a, c'), \ (b, a'), \ (b, b'), \ (b, c'), \ (c, a'), \ (c, b'), \ (c, c'),$$

respectively, by

$$P, \ C', \ B, \ C, \ Q, \ A', \ B', \ A, \ R.$$

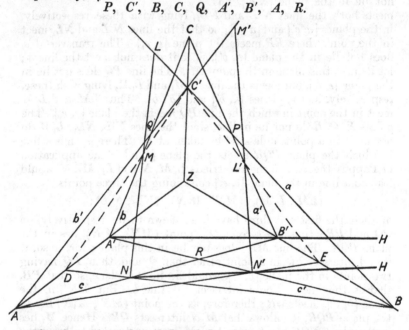

The three planes $[a, a']$, $[b, b']$, $[c, c']$ then meet in pairs in the three lines AA', BB', CC', which thus meet in one point, say Z. Let LMN be a line, meeting the lines a, b, c, respectively, in L, M and N, but not intersecting any of the lines a', b', c'. In the plane $[c, c']$ draw the line BN, meeting the line AA' in the point D. Take N', an arbitrary point on the line c'; in the plane $[c, c']$ draw the line AN', meeting the line BB' in the point E. Let H be the intersection of the lines $A'B'$ and NN', both lying in the plane $[c, c']$. Draw from N' the transversal to meet the lines a' and b', respectively, in L' and M'.

The points *D*, *M*, *C'* are in the plane [*b*, *b'*], and are also in the plane *LBN*, and are therefore in line. The points *C'*, *L'*, *E* are in the plane [*a*, *a'*], and are also in the plane *M'AN'*, and are therefore in line. Thus the line *L'M* lies in the two planes *CA'B'* and *C'DE*, and, therefore, passes through the intersection of the two lines *A'B'* and *DE* of the plane [*c*, *c'*].

Wherefore, the necessary and sufficient condition that the two transversals *LMN* and *L'M'N'* should intersect one another, which is the condition that the lines *L'M* and *NN'* should intersect, is that *DE* should pass through the intersection, *H*, of *A'B'* and *NN'*.

If we assume Pappus' theorem, and apply it to the two triads *A*, *N*, *B'* and *B*, *A'*, *N'*, lying, respectively, on two lines of the plane [*c*, *c'*], the consequence is precisely that *DE* passes through *H*; for *AA'*, *BN* meet in *D*, while *NN'*, *A'B'* meet in *H*, and *AN'*, *BB'* meet in *E*.

Thus condition (2) is satisfied if Pappus' theorem be assumed.

Conversely, from the assumption that the transversals *LMN* and *L'M'N'* intersect one another, the truth of Pappus' theorem follows for the two triads *A*, *N*, *B'* and *B*, *A'*, *N'*. Thus it will be proved that Pappus' theorem follows when condition (2) is satisfied, if we shew that the figure can be constructed when *A*, *N*, *B'* and *B*, *A'*, *N'* are any two sets of three points, each in line, in a plane. For this, take *Z*, the intersection of the lines *AA'* and *BB'*; from *Z* draw the arbitrary line *ZC'C*, not in the plane of the two given triads; join the points *A* and *B* to the arbitrary point, *C'*, of this line, and join the points *A'* and *B'* to the other arbitrary point, *C*, of this line. Then, from the point *N*, draw the transversal *NML* to meet *A'C* in *M* and *BC'* in *L*, and, from the point *N'*, draw the transversal *N'L'M'* to meet *B'C* in *L'* and *AC'* in *M'*. The construction of the points *D*, *E*, *H* then follows as before.

The identification of the two results is thus completed.

It is thus shewn that the adjunction of Pappus' theorem to the Propositions of Incidence enables us to regard the assignment of three points of one line, to correspond to three points of another line, not intersecting the former, as leading to a unique correspondence for all points of two related ranges upon these lines.

Equivalence of Pappus' theorem with the conditions; case of ranges in one plane. We proceed now to shew that the same is true for two related ranges upon lines which intersect one another. That this must be so is evident from what has preceded; for, of two ranges upon intersecting lines, one may be regarded, in an infinite number of ways, as being in perspective with a range lying on a line which does not intersect the other. But an independent proof is interesting, and, incidentally, will furnish information in regard to related ranges in one plane.

Suppose that the ranges $(PQR\ldots)$, $(P'Q'R'\ldots)$, respectively on the intersecting lines AC, AB, are related, being both in perspective with the range $(XYZ\ldots)$, respectively from B_1 and C_1. The line $XYZ\ldots$ must intersect both AC and AB, so that B_1C_1 lies in the

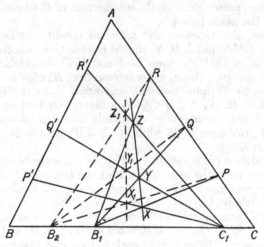

plane of the lines AB, AC, meeting them, respectively, say, in B and C. It can then be shewn that these ranges are related by perspectivities whose centres are any two points on the line BC, as has indeed already appeared in Section I. For take any point, B_2, of BC; let B_2P, B_2Q, B_2R, ... respectively meet C_1P', C_1Q', C_1R', ... in X_1, Y_1, Z_1, ...; then, by Desargues' theorem applied to the two triads PXX_1, PYY_1, since corresponding joins of points of these triads meet in the points B_1, C_1, B_2, which are in line, it follows that X_1Y_1 passes through the point common to XY and AC. The same is true for X_1Z_1; and so on; so that X_1, Y_1, Z_1 are in line. In the same way, if C_2 be another point of BC, and C_2P', C_2Q', C_2R'

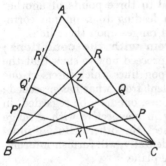

meet B_2P, B_2Q, B_2R respectively in X_2, Y_2, Z_2, we see, from the triads $P'X_1X_2$, $Q'Y_1Y_2$, $R'Z_1Z_2$, that $X_2Y_2Z_2$ are in line.

We may thus suppose the ranges $(PQR\ldots)$ and $(P'Q'R'\ldots)$ to be in perspective with a range $XYZ\ldots$ respectively from the centres B and C.

We first shew that when, of the two related ranges $(PQ\ldots)$, $(P'Q'\ldots)$, the three points P', Q', R' of one are given as corresponding to the three given points P, Q, R of the other, then the point, S', of one range which corresponds to any assigned point, S, of the other, is without ambiguity if we assume Pappus' theorem to be true.

By Pappus' theorem the two triads, P, Q, R and P', Q', R', each of points on a line, give rise to another line containing the three points of intersection $(QR', Q'R)$, $(RP', R'P)$, $(PQ', P'Q)$; and this line is without ambiguity when the two triads are given. But, also by Pappus' theorem, this line in fact is the same as the intermediary line XYZ, arising when the two ranges are related by perspectivities having B, C for centres; for, by applying Pappus' theorem to the two triads P, C, Q and P', B, Q', we infer that the point $(PQ', P'Q)$ lies on the line XY. And we can similarly infer that the points $(QR', Q'R)$ and $(RP', R'P)$ lie on the line XY. But then, if S' be the point of the range $(P'Q'R' \ldots)$ which corresponds to a point S of the range $(PQR \ldots)$, so that BS and CS' meet on XY, a similar proof shews that RS' and $R'S$ meet on XY.

Thus, the rule for the construction of S', when P, Q, R, P', Q', R', and S, are given, is, to find the Pappus line of the two triads P, Q, R and P', Q', R', take the intersection of $R'S$ with this, and draw RS' through this intersection. This construction necessarily gives a unique point, independent of the method whereby the ranges $(PQR \ldots)$, $(P'Q'R' \ldots)$ are related.

The proof of the converse theorem, that the uniqueness of S', when P, Q, R, P', Q', R', and S, are given, requires Pappus' theorem, may be omitted, in view of what has been said above.

Related ranges on the same line. Now consider two ranges on the same line which are related; by definition, this means that one of these ranges is in perspective with a range, on another line, which is related to the remaining range of the given line. Suppose that the two ranges, $(PQRS \ldots)$, $(PQRS' \ldots)$, of the given line have three corresponding points, P, Q, R, in common; and let $(PQRS' \ldots)$ be in perspective, from a centre Z, with a range $(P_1 Q_1 R_1 S_1 \ldots)$, related to $(PQRS \ldots)$, lying on another line, which meets the former in T. Then, by the above, the ranges $(PQRS \ldots)$, $(P_1 Q_1 R_1 S_1 \ldots)$ being related, the joins, RS_1 and $R_1 S$, intersect on the Pappus line of the two triads P, Q, R and P_1, Q_1, R_1, if we assume Pappus'

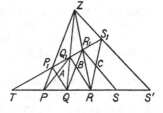

theorem. Denote the points $(PQ_1, P_1 Q)$, $(QR_1, Q_1 R)$, $(RS_1, R_1 S)$, respectively, by A, B, C. By Desargues' theorem, applied to the triads P, A, P_1 and R, B, R_1, the point T lies on AB, which is also the Pappus line of the two triads P, Q, R and P_1, Q_1, R_1. The fact that C lies on this line, involves then, by Desargues' theorem applied to the triads Q, B, Q_1 and S, C, S_1, that the line SS_1 passes through Z. Thus S coincides with S'.

Thus it is a consequence of Pappus' theorem that two related ranges on the same line coincide entirely if they have three corresponding points in common.

Conversely, if two related ranges on the same line coincide entirely when they have three points in common, then Pappus' theorem follows. For, let A, B, C and A', B', C' be any two triads of points respectively on two lines which intersect in X. Consider two related ranges on these lines, so constructed that the points A, B, C, of one of these, correspond respectively to the points A', B', C' of the other; it follows from preceding work that such a construction is possible. Let the point X, of intersection of the lines of the ranges, regarded as a point of the range ($ABC\ldots$), correspond to the point Z' of the second line $A'B'C'\ldots$, in virtue of the perspectivities by which the ranges are related; similarly, let the point X, regarded as belonging to the range ($A'B'C'\ldots$), give rise to the point Y of the line $ABC\ldots$. Join YZ', and let this joining line be met by AA', $A'B$, AB', respectively, in the points P, Q, Q'. The points Y, X, A, B, of the first range, correspond, respectively, to the points X, Z', A', B' of the second range; and, from the centre A', the former range is in perspective with Y, Z', P, Q. Therefore, by what we have proved, the last range is related to the range X, Z', A', B'; this, however, from the centre A, is in perspective with the range Y, Z', P, Q', which is then related to the range Y, Z', P, Q. If, then, we assume that two related ranges on the same line which have three points in common entirely coincide, we can infer that the points Q and Q' coincide, so that the joins AB' and $A'B$ intersect on the line YZ'. By a similar proof the joins AC' and $A'C$ intersect on this line as do the joins BC' and $B'C$. This shews that Pappus' theorem holds for the two triads A, B, C and A', B', C'.

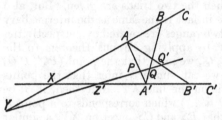

Pappus' theorem and the Principle of Duality. It was seen that the Propositions of Incidence lead to a duality, in virtue of which, in space of three dimensions, points correspond to planes and conversely, and lines to lines. And the conditions (1), (2), (3) above are of self-reciprocal character, since two lines with a common point lie in a plane. As so far stated, Pappus' theorem has not this character, and we must enquire whether its adjunction to the Propositions of Incidence would destroy the validity of the duality of space. We shew that this is not so by deducing the reciprocal of Pappus' theorem from itself and the Propositions of Incidence. In three dimensions this reciprocal would be that, if three planes, α, β, γ, have a line, l, in common, and three planes, α', β', γ', have a line, l', in common, the lines l, l' being in one plane, and we take the lines (α, β'), and (α', β), and the plane containing these, the three planes so obtained have a line in common. By taking a section of the figure by an arbitrary plane, we see that this is equivalent to the statement that, if three lines, a, b, c, lying in a plane, meet in a point, O, and three other lines, a', b', c', of this plane, meet in a point, O', and we denote the points of intersection

$$(b, c'), \ (b', c), \ (c, a'), \ (c', a), \ (a, b'), \ (a', b)$$

respectively by $P, \quad P', \quad Q, \quad Q', \quad R, \quad R',$

then the three lines PP', QQ', RR' meet in a point.

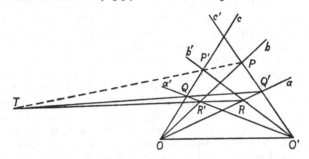

In fact, if T be the point of intersection of the lines QQ' and RR', the application of Pappus' theorem to the two triads, O, Q', R and O', R', Q, shews that the line PP' passes through T.

Related ranges in space of more than three dimensions. If in space of n dimensions we have a succession of ranges of which every consecutive two (which will both lie in a threefold space) are related by the definition applied before, the proof given in Section I of this Chapter avails to shew that the first and last are related, and assuming Pappus' theorem to hold generally, it follows as here that three points are sufficient to identify a range related to a given one. This leaves open the question, suggested in the examples in Section I, whether a more compendious method of relating two ranges in space of higher dimensions may not be possible than by a succession of perspectivities by lines drawn from a centre. The next Section (III) throws some light on the general question here touched. For the present we shall only add the further remark that, as in space of three dimensions, the adjunction of Pappus' theorem leads to no abandonment of the Principle of Duality, the reciprocal of Pappus' theorem in space of n dimensions being, as in that case, reducible to the theorem itself. The reciprocal in question would be as follows : In space of n dimensions we are given two spaces, each of $n-2$ dimensions, which we shall denote by λ and λ', both containing the same space S_{n-3}, say ϖ (this being the reciprocal of the plane in which the lines l, l' of the original theorem both lie). Through the S_{n-2} called λ there pass then three spaces S_{n-1}, say U, V, W, and, similarly, three spaces S_{n-1}, say U', V', W', through λ'. Then the spaces U, V' intersect in a space S_{n-2}, as do also the spaces U', V; and the two S_{n-2} so obtained both lie in a S_{n-1}, which we call w. The pairs U, W' and U', W similarly determine a S_{n-1}, which we call v; and the pairs V, W' and V', W determine a S_{n-1}, which we call u. The spaces $U, V, W, U', V', W', u, v, w$ all contain the S_{n-3} which we call ϖ. The theorem is that the spaces u, v, w have a space S_{n-2} in common.

To see that this is true, take a general plane, σ, in the space of n dimensions, not intersecting the space S_{n-3} which we have called ϖ. This plane, σ, will be met by λ, λ' each in a point, say in the points O, O', respectively ; it will be met by U, V, W in lines through O, say a, b, c respectively, and by U', V', W' in lines, a', b', c', through O'. The space w then meets the plane σ in the line joining the points (a, b') and (a', b), and similarly with the spaces u and v. We therefore have the same figure in the plane σ as was obtained in three dimensions by reciprocating the figure of Pappus' theorem. Thus the lines in the plane σ arising from the spaces u, v, w meet in a point. This point with the space ϖ, of $n-3$ dimensions, which is contained in each of the spaces u, v, w, determine a space S_{n-2} contained in all of these. This constitutes the space which proves the theorem.

Examples in connexion with Pappus' theorem.

Ex. 1. Pappus' theorem can be *proved* for the case of two triads which are in perspective.

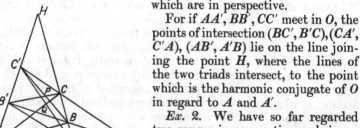

For if AA', BB', CC' meet in O, the points of intersection $(BC', B'C), (CA', C'A), (AB', A'B)$ lie on the line joining the point H, where the lines of the two triads intersect, to the point which is the harmonic conjugate of O in regard to A and A'.

Ex. 2. We have so far regarded two ranges in perspective as being a particular case of two related ranges, of which the two centres of perspective coincide, and the intermediary line is any line through the point of intersection of the lines of the two ranges.

There are however other more definite ways in which two ranges in perspective may be related, with two distinct centres of perspective.

(*a*) Two ranges in perspective may be related with the inter-mediary line passing through the point of intersection of the lines of the ranges and the sup-plementary line, that is the line joining the two centres of per-spective, passing through the original centre of perspective.

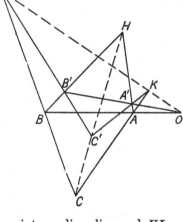

For if the ranges (*AA'* ...), (*BB'* ...) be in perspective from *O*, and *K*, *L* be any two points on a line through *O*, the lines *KA*, *LB* meeting in *C*, and the lines *KA'*, *LB'* meeting in *C'*, it follows, by Desargues' theo-rem, that *C'* lies on the line joining *C* to the point *H*, the point of intersection of the lines of the ranges. The original ranges are thus related with *CH* as intermediary line and *KL* as supplementary line.

(*b*) If we assume Pappus' theorem we can relate the ranges in per-spective in such a way that, not the intermediary line, *CH*, but, the supplementary line, *KL*, passes through the intersection *H* of the lines of the ranges.

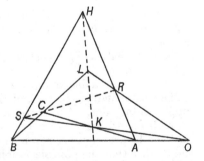

For let the ranges (*A* ...), (*B* ...), on lines intersecting in *H*, be in perspective from *O*, and *HLK* be an arbitrary line through *H*, while *RS* is an arbitrary line meeting *HA, HB*, respectively, in *R* and *S*, not passing through *O* or *H*; let *OR, OS* meet the line *HLK* respec-tively in *L* and *K*. Then, using Pappus' theorem for the two triads *A*, *O*, *B* and *L*, *H*, *K*, we see that *AK* and *BL* meet on *RS*. Thus the two ranges are in perspective with a range on *RS*, respectively from the centres *K* and *L*; the line *RS* is now the intermediary line, and the line *KL* the supplementary line.

(*c*) Conversely, two related ranges are in perspective with one another when the intermediary line passes through the point of intersection of the lines of the ranges; and the new centre of perspective is on the supplementary line of the ranges. This follows by Desargues' theorem. The diagram in (*a*) may be used.

(*d*) And two related ranges are also in perspective with one another when the supplementary line passes through the point of intersection of the lines of the ranges. This follows from Pappus' theorem. The diagram in (*b*) may be used.

(*e*) It is clear that if, of two related ranges, the point of intersection of the lines of the ranges corresponds to itself regarded as belonging to both lines, then either the intermediary line or the supplementary line passes through this point. Hence, with the assumption of Pappus' theorem, it follows, from what has been said, that the two ranges are in perspective with one another.

Ex. 3. We have seen, assuming Pappus' theorem, that, in the

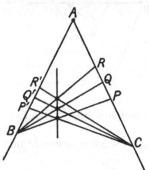

case of two related ranges on intersecting lines for which the supplementary line does not pass through the point of intersection of the lines of the ranges, the two centres of perspective, by means of which the ranges are related, can be taken to be any two points on the supplementary line. In particular if these centres, B and C, be respectively on the lines, $(P', Q', ...)$ and $(P, Q, ...)$, of the two ranges, then the two triads B, P', Q' and C, P, Q have the same Pappus line as have the two triads P, Q, R and P', Q', R', this being the intermediary line for the two related ranges.

Thus the centres of perspective, B and C, are corresponding points of the two ranges.

More generally a line joining any point, S', of the range $(P', Q', ...)$, to the corresponding point, S, of the range $(P, Q, ...)$, is a possible supplementary line. And it will appear later (in Volume II) that, through an arbitrary point of the plane of the two ranges, two lines can be drawn either of which may serve as supplementary line; and further that any possible supplementary line for the two ranges is intersected, by the joins $PP', QQ', ...$, of corresponding points of the two ranges, in a range of points which is related to either of the two given ranges. This last is the same result as was found for related ranges on two lines which do not intersect, before the introduction of Pappus' theorem.

Ex. 4. Consider two related ranges $(A, B, C, D, ...)$ and $(A', B', C', D', ...)$, on two lines which do not intersect one another. For each pair of corresponding points of these ranges, such as A, B and

A', B', consider the cross joins (not joining corresponding points), such as AB' and $A'B$. Let O be any point of the threefold space of the two ranges; from this point a transversal can be drawn to the cross joins AB' and $A'B$. It can be shewn that all such transversals lie in a plane through O; such a plane is then determined, by the two given ranges, passing through every point O. Further, if the transversal from O to the lines of the two given ranges meet them, respectively, in P and Q', and P', Q be the points of the ranges $(A'B' ...)$, $(AB ...)$ respectively corresponding to P, Q', this plane is that joining O to the line $P'Q$.

This result, which is of subsequent interest, is a consequence of Pappus' theorem. Taking an arbitrary plane, not passing through O, the given ranges are projected from O upon this plane into two ranges, $(A_1, B_1, C_1, ...)$ and $(A_1', B_1', C_1', ...)$, related to one another, for which, as we have shewn to follow from Pappus' theorem, the points of intersection $(A_1B_1', A_1'B_1)$, $(A_1C_1', A_1'C_1)$, $(B_1C_1', B_1'C_1)$, ..., are in line.

In particular, when P, Q' and P', Q are constructed, as in the enunciation, instead of one there is an infinite number of transversals from O to the cross joins PQ', $P'Q$, namely any line in the plane $OP'Q$ is such a transversal.

Ex. 5. The reciprocal theorem is that, if $(A, B, C, ...)$ and $(A', B', C', ...)$ be related ranges on two lines, l and l', which do not intersect one another, and if an arbitrary plane, ϖ, meet these lines, l and l', respectively in P and Q', to which there correspond, on the other lines, l' and l, respectively, the points P' and Q, then the point O, in which the plane ϖ is met by the line $P'Q$, lies on the line joining the intersections of the plane ϖ with any pair of cross joins, such as AB' and $A'B$. By the two given ranges there is thus determined a point, O, upon any plane, ϖ.

Ex. 6. If A, B, C, A', B', C' be six general points in threefold space, and O be any point on the line of intersection of the planes ABC and $A'B'C'$, and we draw from O the transversal to the pair of joins BC' and $B'C$, as also the transversal to CA' and $C'A$, and the transversal to AB' and $A'B$, these three transversals are in one plane.

Ex. 7. If, in fourfold space, the four points A, B, C, D be in one plane, and the four points A', B', C', D' be in another plane, the three transversals, respectively, of the triplets of lines $(BC', B'C, DD')$, $(CA', C'A, DD')$, $(AB', A'B, DD')$, are in one threefold space. This space then contains the line DD'; if we project the figure on to an arbitrary plane by means of planes passing through the line DD', we obtain Pappus' theorem.

Ex. 8. In Pappus' theorem for two triads, A, B, C and A', B', C', there are six ways of arranging the triad A', B', C' in association with A, B, C. Consider the figure of the six Pappus lines thus found. And reciprocate the figure.

Ex. 9. The following result is of great importance for the development of the theory.

We have seen that if five arbitrary points be taken on a line,

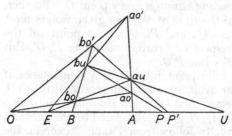

consisting of two couples, O, U and A, B, and a determining point E, and there be drawn, in an arbitrary plane passing through this line, lines o, u, a, b respectively through the points O, U, A, B, subject to the condition that the intersections, (a, u) and (b, o), lie on a line through the point E, then the join of the points, (a, o) and (b, u), meets the original line in a point, P, independent of the lines o, u, a, b. Similarly, if the lines o, u, a, b be subject to the condition that the intersections, (a, o) and (b, u), lie on a line through E, then the join of the points, (a, u) and (b, o), meets the original line in a point, P', also independent of the lines o, u, a, b.

It was remarked that the Propositions of Incidence are insufficient to determine whether P and P' coincide. It is important to notice that, if Pappus' theorem be assumed, these points must be regarded as identical.

For we can have a figure containing both constructions, the lines a, b, u being the same in both, the line drawn through O for the second construction being o'. If then, on the lines a, b, we consider the two triads of points, respectively

$$(a, u), \quad (a, o'), \quad (a, o),$$
and $$(b, u), \quad (b, o), \quad (b, o'),$$

then the first and second of these three pairs of corresponding points, by the joins of the points, (a, u) to (b, o) and (a, o') to (b, u), give the point E; the second and third, by the joins of (a, o) to (b, o) and of (a, o') to (b, o'), give the point O; but the first and third, of (a, o) to (b, u) and of (a, u) to (b, o'), are the lines respectively containing P and P'. These lines thus meet the original line, EO, in the same point.

In the original figure, before the consideration of Pappus' theorem, the points O, U, of a couple, were symmetrical, as were also the points A, B of the other couple. The introduction of Pappus' theorem renders the points E, P symmetrical in regard to one another. This is brought out well by a diagram illustrating the two cases in which lines o, u, a, b are drawn in one plane, and other lines o', u', a', b', through the same points, are drawn in another plane through the original line. The diagram so obtained is precisely that which, with different lettering, is used to illustrate the following example.

Ex. 10. Let A, B, C, D be any four points not in one plane; let A', B', C', D' be further points lying, respectively, in the planes BCD, CAD, ABD, ABC; suppose that the line $B'C'$ meets AD', the line $C'A'$ meets BD' and the line $A'B'$ meets CD'. Thus the planes $B'C'D'$, $C'A'D'$, $A'B'D'$ contain, respectively, the points A, B, C. It is supposed that the points A', B', C' are not in the plane

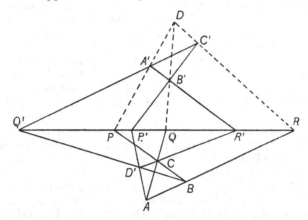

ABC (respectively on the lines BC, CA, AB), so that the points A', B', C', D' are not in one plane.

We can then shew, assuming Pappus' theorem, that the plane $A'B'C'$ contains the point D.

Consider the line of intersection of the two planes ABC and $A'B'C'$. As $B'C'$, AD' meet, their intersection, P', must be on this line; as must similarly be the point of intersection, Q', of $C'A'$, BD', and the intersection, R', of $A'B'$, CD'. Let BC, CA, AB meet this line, respectively, in P, Q, R.

Then we have a figure such as that spoken of in the last example. Considering the plane of the points A, B, C, D', and denoting by p, p' and r, r' the lines drawn in that plane respectively through the points P, P', R, R', the join of the points (p, r') and (p', r) passes through the point Q, and the join of the points (p, r) and (p', r') passes through the point Q'. In the other plane, consider the lines PA' and RC', denoting these, respectively, by p_1 and r_1, the lines drawn in this plane through P' and R' being respectively denoted by p_1' and r_1'. It follows then, from the fact that the join of the points (p_1, r_1') and (p_1', r_1) passes through Q', that the join of the points (p_1, r_1) and (p_1', r_1') passes through Q. Thus the lines PA', QB', RC' meet in one point. This point is evidently on each of the planes $A'BC$, $B'CA$, $C'AB$, and is therefore the point D. It is thus proved that the plane $A'B'C'$ contains the point D.

It is thus shewn that it is possible to have two tetrads of points A, B, C, D and A', B', C' D' with the relation that the plane containing any three points of either tetrad contains a point of the other tetrad. This remark was first made by Moebius, and we shall therefore sometimes refer to the tetrads as Moebius tetrads.

It is also shewn incidentally that the three pairs of opposite joins of four points in a plane meet an arbitrary line of that plane in six points of which one·is to be regarded as determined uniquely by the other five. It is usually said that the three couples, of which the six points are composed, are three couples in *involution*. The relation was recognised by Pappus (cf. Ch. Taylor, *Ancient and Modern Geometry of Conics*, Cambridge, 1881, p. lii). It is important for our purpose to notice that it does not follow from the Propositions of Incidence, without what we have called Pappus' theorem.

The theorem of Moebius tetrads may be stated thus : Let P_1, P_2, P_3, P_4 be four arbitrary points in a plane ; through each of the six joins of two of these let an arbitrary plane be drawn, the plane through $P_i P_j$ being denoted by α_{ij}. The set of three of these planes, passing through the pairs from three of the original points, meet in a point; for instance the planes α_{23}, α_{31}, α_{12} meet in a point which we may call P_{123}. The theorem is then that the four points P_{234}, P_{314}, P_{124}, P_{123} lie in one.plane.

The reciprocal theorem, of which the generalisation to space of higher dimensions will be of interest later, is as follows : Let four planes α_1, α_2, α_3, α_4 be drawn through a point O ; on the line of intersection of any two of these planes take an arbitrary point, say, the point P_{ij} on the line (α_i, α_j) ; through the three such points arising from three of the planes there passes a plane, for instance the plane α_{123} through the points P_{23}, P_{31}, P_{12}. The four planes α_{234}, α_{314}, α_{124}, α_{123} meet in a point, say O'. In all we have eight points and eight planes ; through each point pass four of the planes, and in each plane lie four of the points.

SECTION III. INTRODUCTION OF ALGEBRAIC SYMBOLS

The laws of operation of the symbols employed. We have considered in the two preceding Sections, respectively, the Propositions of Incidence, and the consequences of supposing Pappus' theorem to hold. It will add greatly to clearness, and, finally, will amply justify itself as a means of investigation, to accompany the preceding work by an algebraic symbolism. And as an example of one kind of usefulness which this may have, it may be as well to cite at once the following, which signalises an important step in the logical development. It may perhaps be supposed that as Desargues' theorem in a plane was deduced in consequence of the Propositions of Incidence for space of three dimensions, so it may be possible to obtain Pappus' theorem as a consequence of Propositions of Incidence assumed for higher space. It will appear however that there is an algebraic symbolism corresponding in every detail with the geometry deducible from the Propositions of Incidence alone, in space of however high dimensions, which is sharply and definitely distinguished from the symbolism which is appropriate if Pappus' theorem is true. This theorem requires a precise limitation, in the character of the symbols adequate when only the Propositions of Incidence are assumed; thus Pappus' theorem cannot be a consequence of Propositions of Incidence in however high dimensions.

The symbols which we first introduce are, speaking in general terms, subject to all the laws of ordinary algebra, except the commutative law of multiplication. They are not necessarily, and will not be finally, arranged in order of magnitude, so that, in their entirety, they are wider than the real numbers of arithmetic. In order to emphasize this, we give here a formal statement of the properties assumed for them, and some examples.

We use the symbol = to mean " may be replaced by." Thus if a, b be two of the symbols, the statement $a = b$ is the same as $b = a$.

From any two of the symbols, say a and b, another symbol can be formed, represented by $a + b$, or by $b + a$, independent of the order in which a, b are taken. Further, if $a = b$ and $x = y$, then also $a + x = b + y$. Also, if a, b, c be any three of the symbols, the symbol thus formed from a and $b + c$ is equivalent with that so formed from $a + b$ and c; that is

$$a + (b + c) = (a + b) + c.$$

We speak of the process of forming $a + b$ as *addition*, and of $a + b$ as the *sum* of a and b; and we may say that the addition of two symbols is *commutative*, and the addition of three symbols is *associative*. It follows that the symbol

$$a + b + c + \ldots + k,$$

formed from any number of the symbols, has a definite meaning independent of the order, and of the partial associations, of the symbols.

Among the symbols however there is one, and only one, which

we denote by 0, whose addition to any other symbol leaves that unaltered, so that

$$a + 0 = 0 + a = a.$$

To every symbol, say a, there corresponds another, which we denote by $-a$, whose addition to the original gives 0 for sum, so that

$$a + (-a) = 0 = (-a) + a.$$

The symbols so corresponding to any two equivalent symbols are themselves equivalent, so that if $a = b$, then also $-a = -b$. Thus the symbol so corresponding to 0 is itself 0; for -0 is, by the definition of 0, equal to $0 + (-0)$, and this, by the definition of -0, is itself 0. Conversely, if a be such that

$$a = -a,$$

it follows that $a + a = 0$; it will be a consequence of subsequent prescriptions that thence $a = 0$. We denote the sum of two symbols a and $-b$, namely $-b + a$, or $a + (-b)$, also by $a - b$. Thus if $a = b$ and $x = y$ we also have $a - x = b - y$. And, by the associative law, whatever symbols a and b may be, we have

$$(a - b) + b = a + (-b) + b = a + (-b + b) = a + 0 = a,$$
$$a - (a - b) = [(a - b) + b] - (a - b) = b + (a - b) - (a - b) = b.$$

Thus also, if $b = -a$, then $-b = a$, for

$$-b = 0 - b = a + (-a) - b = a + b - b = a,$$

and either of these follows from $a + b = 0$. We may speak of $-a$ as the negative of a.

From any two symbols, a, b, can be formed, beside the sum $a + b, = b + a$, two further combinations, different from this, dependent on the order in which a, b are considered, denoted, respectively, by

$$a \cdot b \text{ or } ab, \text{ and } b \cdot a \text{ or } ba.$$

It may happen that these symbols are equivalent, but this is not to be assumed in general. In particular, for the symbol 0, whatever a may be, we have $0 \cdot a = a \cdot 0 = 0$.

If $a = b$ and $x = y$, then $ax = by$, and $xa = yb$. Also, if three symbols a, b, c be taken, the consequence of combining, in this way, the symbol ab, taken first, with c, is the same as of combining a, taken first, with bc; namely

$$(ab) c = a (bc).$$

We may speak of ab as obtained by *multiplication* of a and b, and of ab as their *product*, and say that multiplication is *associative*. Thus if $a, b, ..., k$ be any symbols, the symbol

$$ab ... k$$

is in general dependent upon the order of the component symbols, but not dependent upon their association to form partial products, being the same, for instance, as

$$(ab)(c \ldots k), \text{ or } a(bc \ldots k).$$

We may also have multiplications $(a + b)c$, or $c(a + b)$, defined respectively by

$$(a + b)c = ac + bc, \quad c(a + b) = ca + cb.$$

This leads, if a, b be any symbols, to the result

$$(-a)b = -(ab) = a(-b);$$

for $\qquad (-a)b + ab = (-a + a)b = 0 \cdot b = 0,$

$$a(-b) + ab = a(-b + b) = a \cdot 0 = 0.$$

Beside the special symbol 0, with the two properties

$$a + 0 = b + a = a, \quad a \cdot 0 = 0 \cdot a = 0,$$

there is another special symbol, denoted by 1, with the property that, whatever symbol a may be,

$$a \cdot 1 = 1 \cdot a = a.$$

For reasons which will appear, the symbol 0 is said to be *singular*. There may be other singular symbols. The product of two non-singular symbols is not singular.

To every symbol, a, which is not singular, there corresponds another symbol, denoted by a^{-1}, having the property that

$$a^{-1} \cdot a = a \cdot a^{-1} = 1,$$

and such that, if $a = b$, then $a^{-1} = b^{-1}$. Thus

$$ab^{-1} \cdot b = a \cdot b^{-1}b = a \cdot 1 = a, \quad a \cdot a^{-1}b = aa^{-1} \cdot b = 1 \cdot b = b,$$

while, if a, b be symbols such that $ab = 1$, and a be not singular,

$$a^{-1} = a^{-1} \cdot 1 = a^{-1}(ab) = a^{-1}a \cdot b = b,$$

from which, if also b be not singular,

$$b^{-1} = 1 \cdot b^{-1} = aa^{-1} \cdot b^{-1} = ab \cdot b^{-1} = a \cdot bb^{-1} = a.$$

Further, if $x = ab$, then $x^{-1} = b^{-1}a^{-1}$; for, if $y = b^{-1}a^{-1}$,

$$xb^{-1} = a, \quad xb^{-1}a^{-1} = 1, \quad xy = 1, \quad y = x^{-1}.$$

Also, if a is not singular, there is one and only one symbol, x, such that $ax = b$, and one and only one symbol y such that $ya = b$, where b is an arbitrary symbol. For $ax = b$, $ya = b$ lead, respectively, to

$$x = a^{-1}a \cdot x = a^{-1} \cdot ax = a^{-1}b, \quad y = y \cdot aa^{-1} = ya \cdot a^{-1} = ba^{-1}.$$

This involves that the statement $ax = az$ requires $x = z$, the symbol a not being singular.

It is to be remarked that the symbol 1 is by no means necessarily identical with that so denoted in ordinary arithmetic, as subsequent examples will illustrate. But no confusion need arise, since the laws of combination agree. Similarly we may use the symbol 2 for the sum $1 + 1$, and 3 for the sum $1 + 1 + 1$, and so on. It may then be shewn, as in ordinary arithmetic, that these combine together just as do the symbols of ordinary arithmetic—and in particular that they are commutative in multiplication. For distinctness we shall speak of them as the *iterative* symbols. None of these symbols is singular, and to any one of them, for instance for 2, or for 3, there is a symbol 2^{-1}, or 3^{-1}, of which 1^{-1} is equivalent to 1 ; if m, n be any two of these symbols, we have $m^{-1}n = nm^{-1}$, this being equivalent to $nm = mn$. As regards combinations with other of our symbols, we have

$$2 \cdot a = (1+1)\, a = a + a = a(1+1) = a \cdot 2,$$
$$2a \cdot b = (a + a)\, b = ab + ab = 2 \cdot ab = a(b + b)$$
$$= a \cdot 2b = a2 \cdot b = a \cdot b \cdot 2 = ab \cdot 2,$$

and so on; so that if, in a product $abc \ldots k$, any of the symbols a, b, ..., k be those denoted by 1, 2, 3, ..., the places where they occur in the product are indifferent, so long as the order of the other symbols is preserved.

Examples of symbols obeying these laws. Simple examples of such symbols may be formed with aggregates of the numbers of ordinary arithmetic, the addition and multiplication of the symbols being defined by arithmetical operations carried out, according to a prescribed rule, with the numbers of the aggregate. If a, a', β, β', ... be numbers of ordinary arithmetic, including negative and fractional numbers, the numbers zero and unity being, for distinctness, denoted by $\bar{0}$, $\bar{1}$, we may, for example, have symbols formed with two such numbers, taken in a definite order,

$$a = (a,\ a'), \quad b = (\beta,\ \beta'),\ \ldots,$$

defining the fundamental symbols and operations by

$$0 = (\bar{0},\ \bar{0}), \quad 1 = (\bar{1},\ \bar{0}), \quad a + b = (a + \beta,\ a' + \beta'), \quad -a = (-a,\ -a'),$$
$$ab = (a\beta - a'\beta',\ a\beta' + a'\beta),$$
$$a^{-1} = \left(\frac{a}{a^2 + a'^2},\ \frac{-a'}{a^2 + a'^2} \right).$$

Here 0 is the only singular symbol, and every two symbols are commutative in multiplication.

Another example is furnished by symbols which are aggregates of three ordinary numbers, taken in a definite order, say

$$a = (a,\ a',\ a''), \quad b = (\beta,\ \beta',\ \beta''),\ \ldots,$$

with
$$0 = (\bar{0},\ \bar{0},\ \bar{0}), \quad 1 = (\bar{1},\ \bar{0},\ \bar{1}), \quad a + b = (a + \beta,\ a' + \beta',\ a'' + \beta''),$$
$$-a = (-a,\ -a',\ -a''), \quad ab = (a\beta,\ a\beta' + a'\beta'',\ a''\beta''),$$
$$a^{-1} = \left(\frac{1}{a},\ -\frac{a'}{aa''},\ \frac{1}{a''} \right).$$

Here every symbol (a, a', a'') for which either a or a'' vanishes is a singular symbol; and the equation $ab=ba$ is not generally true.

A very general example of such symbols as we have described is furnished by those aggregates of n^2 ordinary numbers which are called matrices; here n may be two, or three, or any positive integer number. The n^2 component numbers are thought of as arranged in n rows and n columns, so that it is convenient to denote the number in the rth row and sth column by $a_{r,s}$. Two such symbols are considered equivalent only when they have the same number of component numbers, and these are separately the same. The sum of two such symbols, a, b, of the same number of components, generally denoted by $a_{r,s}$ and $\beta_{r,s}$, is the symbol whose general component is $a_{r,s}+\beta_{r,s}$; the symbol 0 is that of which every component vanishes; the symbol $-a$ is that of which the general component is $-a_{r,s}$. The symbol ab is that of which the general component, $\gamma_{r,s}$, is formed, from the rth row of a, and the sth column of b, according to the rule

$$\gamma_{r,s}=a_{r,1}\beta_{1,s}+a_{r,2}\beta_{2,s}+\ldots+a_{r,n}\beta_{n,s},$$

where, in the suffixes, the numbers 1, 2, ... are those of ordinary arithmetic. It is easy to prove from this definition that, for three such symbols a, b, c, with each n^2 components, we have

$$ab \cdot c=a \cdot bc$$
$$(a+b)c=ac+bc, \quad c(a+b)=ca+cb,$$

but not in general $ab=ba$. The symbol 1, when the number n^2 is specified, denotes the matrix in which every component $a_{r,s}$ is zero except the diagonal components $a_{r,r}$, of which each is unity. The symbol 2 may then denote that formed by an equal number of components, of which $a_{r,s}=0$, except $a_{r,r}$, which is the number 2 of ordinary arithmetic, and so on. It will easily be seen that this is in accordance with the general specifications above, this description being still applicable when 2 is replaced by any number of ordinary arithmetic. In particular, if h be any number of ordinary arithmetic, and a be a matrix symbol, say of n^2 components $a_{r,s}$, it will be legitimate to regard the matrix whose general component is $ha_{r,s}$ as the product of a and the matrix denoted by h. To define the symbol a^{-1}, denote by $|a|$ the determinant formed with the n^2 components, and by $A_{r,s}$ the cofactor of $a_{r,s}$ therein, with its proper sign. Then a^{-1} is the matrix of n^2 components of which the component in the rth row and sth column is given by

$$a'_{r,s}=A_{s,r}/|a|.$$

When the components are such that the determinant $|a|$ is zero, there is no symbol a^{-1}. In this case a is a singular symbol; and there may evidently, for specified n, be an infinite number of such symbols.

Particular examples of such matrix symbols may be referred to:

(1) For $n=2$, if a, a' be numbers of ordinary arithmetic, of which the numbers zero and one are denoted by $\bar{0}$, $\bar{1}$, and we put

$$1=\begin{pmatrix}\bar{1}, & \bar{0}\\ \bar{0}, & \bar{1}\end{pmatrix}, \quad i=\begin{pmatrix}\bar{0}, & -\bar{1}\\ \bar{1}, & \bar{0}\end{pmatrix},$$

then the symbol
$$a=\begin{pmatrix}a, & -a'\\ a', & a\end{pmatrix},$$

may be replaced by

$$a\begin{pmatrix}\bar{1}, & \bar{0}\\ \bar{0}, & \bar{1}\end{pmatrix}+a'\begin{pmatrix}\bar{0}, & -\bar{1}\\ \bar{1}, & \bar{0}\end{pmatrix}, \quad =a \cdot 1+a' \cdot i;$$

we find
$$1 \cdot 1=1, \quad 1 \cdot i=i \cdot 1, \quad i \cdot i=-1,$$

5—2

and hence

$$ab=(a.1+a'.i)(\beta.1+\beta'.i)=(a\beta-a'\beta')1+(a\beta'+a'\beta)i$$
$$=\begin{pmatrix} a\beta'-a'\beta', & -(a\beta'+a'\beta) \\ a\beta'+a'\beta, & a\beta'-a'\beta \end{pmatrix},$$

so that $\qquad\qquad ab=ba.$

If, for brevity, m stand for $a^2+a'^2$, the symbol a^{-1} is given by

$$a^{-1}=\begin{pmatrix} \dfrac{a}{m}, & \dfrac{a'}{m} \\ -\dfrac{a'}{m}, & \dfrac{a}{m} \end{pmatrix},$$

which is impossible if $a=0$, $a'=0$.

The symbols here described have precisely the same behaviour as the independent variables of ordinary analysis. They are therefore of the highest importance.

(2) If we similarly consider symbols given by

$$a=\begin{pmatrix} a, & a' \\ \bar{0}, & a'' \end{pmatrix}, \quad b=\begin{pmatrix} \beta, & \beta' \\ \bar{0}, & \beta'' \end{pmatrix}, \ \dots$$

where $\bar{0}$ is the zero and $a, a', \dots, \beta, \dots$ are numbers of ordinary arithmetic, we find

$$ab=\begin{pmatrix} a\beta, & a\beta'+a'\beta'' \\ \bar{0}, & a''\beta'' \end{pmatrix}, \quad a^{-1}=\begin{pmatrix} \dfrac{1}{a}, & -\dfrac{a'}{aa''} \\ \bar{0}, & \dfrac{1}{a''} \end{pmatrix},$$

so that ab is different from ba in general, and a^{-1} is impossible if either a or a'' be zero.

(3) As the symbols considered in Ex. (1) obey all the laws of combination of the numbers of ordinary arithmetic, they may themselves be used as components to define other symbols. Using $0, 1, i$ for the particular symbols of this kind defined in Ex. (1), let us consider then the four symbols

$$U=\begin{pmatrix} 1, & 0 \\ 0, & 1 \end{pmatrix}, \quad I=\begin{pmatrix} 0, & -1 \\ 1, & 0 \end{pmatrix}, \quad J=\begin{pmatrix} 0, & i \\ i, & 0 \end{pmatrix}, \quad K=\begin{pmatrix} -i, & 0 \\ 0, & i \end{pmatrix};$$

it is then easy to compute that

$$JK=-KJ=I, \quad KI=-IK=J, \quad IJ=-JI=K,$$
$$I^2=J^2=K^2=-U.$$

Hence, if a, β, γ, δ be iterative symbols, that is, here, matrices of two rows and columns with components zero except in the diagonal where there are equal numbers of ordinary arithmetic—so that, in brief, a, β, γ, δ behave like numbers of ordinary arithmetic—, the symbol

$$a=\delta+aI+\beta J+\gamma K$$

is the same as $\qquad a=\begin{pmatrix} \delta-\gamma i, & -a+\beta i \\ a+\beta i, & \delta+\gamma i \end{pmatrix}.$

Replacing a, β, γ, δ respectively by $a', \beta', \gamma', \delta'$, to form a symbol a', we easily find that aa' is generally different from $a'a$. In particular, the symbol a^{-1} is obtained by replacing a, β, γ, δ respectively by

$$a'=-m^{-1}a, \quad \beta'=-m^{-1}\beta, \quad \gamma'=-m^{-1}\gamma, \quad \delta'=m^{-1}\delta,$$

where m denotes $a^2 + \beta^2 + \gamma^2 + \delta^2$. Thus a^{-1} is impossible when $a=0$, $\beta=0$, $\gamma=0$, $\delta=0$, in which case $a=0$.

Another symbol with the same laws of combination as a may be represented by

$$\begin{pmatrix} \delta, & \gamma, & -a, & -\beta \\ -\gamma, & \delta, & \beta, & -a \\ a, & -\beta, & \delta, & -\gamma \\ \beta, & a, & \gamma, & \delta \end{pmatrix}.$$

Beside the symbols formed with aggregates of ordinary numbers, there are other symbols obeying the laws of combination which we have enunciated. Of such those introduced by Grassman are probably the most important. For a definite positive integer number n, consider symbols $e_1, e_2, \ldots, e_n$ such that

$$e_i^2 = \bar{0}, \quad e_i e_j = -e_j e_i, \qquad (i, j = 1, 2, \ldots, n).$$

With these we can form a symbol

$$a = a + a_1 e_1 + \ldots + a_n e_n,$$

where $a, a_1, \ldots, a_n$ are (in the first instance) numbers of ordinary arithmetic. The symbols 0, 1 are obtained then by putting, respectively,

$$a = a_1 = \ldots = a_n = \bar{0}, \quad \text{and} \quad a - \bar{1} = a_1 = \ldots = a_n = \bar{0};$$

while a^{-1} is given by

$$a^{-1} = \frac{1}{a} - \frac{a_1}{a^2} e_1 - \ldots - \frac{a_n}{a^2} e_n,$$

and is impossible, whatever $a_1, \ldots, a_n$ may be, when $a=0$.

Preliminary remarks in regard to the use of the symbols in geometry. Some brief remarks seem necessary to make clear the point of view taken at this stage in introducing the symbols into geometrical discussions. It appears to be more interesting, as well as more satisfactory, to regard this introduction at present as provisional, and illustrative, and to require, for every geometrical result, a geometrical proof independent of the symbols. In the first place we shew that the Propositions of Incidence, and the consequences of these, can be well represented by the symbols. It is more difficult to be sure that every result obtainable from the symbolism, when interpreted as a geometrical theorem, is demonstrable by geometrical reasoning; but we give in detail a geometrical representation of every one of the fundamental laws of combination of the symbols, in accordance with the Propositions of Incidence.

We then shew that the introduction of Pappus' theorem corresponds to a definite additional limiting law of combination for the symbols, namely that their multiplication is commutative. If the correspondence between the geometry and the symbols be exact, it will then follow that a geometrical result is a consequence of the Propositions of Incidence only, when its representation and deduction by means of the symbols does not require this commutative property. In attempting to construct a geometrical proof of a suspected theorem, this remark is of value, and we give several examples of its application.

Two kinds of symbols are employed in what follows: Symbols for geometrical elements, generally points, which we may call element-symbols; and symbols subject to laws of computation, which we may call algebraic symbols. It is of these last we have so far spoken in what precedes. In dealing with any geometrical figure we have as many fundamental element-symbols as is necessary to fix the symbols belonging to all the points, or elements, of the figure, together with a certain number of algebraic symbols. From the manner in which the symbols are introduced it will appear to be without doubt that two points which are identical in virtue of their geometrical construction must then have, as a consequence of computation, the same element-symbol. If we are to employ the symbolism with confidence the converse must also be established, the identity of two element-symbols, arrived at by computation, must imply the possibility of proving the identity of the elements by geometrical reasoning; we seek to induce the conviction that this is so in the present Section, reserving a final consideration for Chapter III.

The use of algebraic symbols for geometrical reasoning dates from very early times (? Apollonius B.C. 230; Ptolemy A.D. 130), and has prompted much in the development of Abstract Geometry. To some writers, therefore, it has seemed to need no justification. But first, it seems evident that a geometry prompted by an algebra must necessarily run parallel to the laws of combination of the symbols of that algebra; and, as we have sought to indicate, there is a good deal of arbitrariness in such laws. And, second, it would seem that a geometry, properly so called, must necessarily be based upon an analysis of our intuition, or experience, of geometrical relations; and that it is necessary to shew that the ideas we adopt from this experience, are properly represented by the symbols. It does not seem likely that a purely analytical geometry can convey any geometrical ideas; there must be constant appeal to geometrical conceptions to give it a meaning; it is surely proper that these conceptions should receive consideration before the symbols are introduced. And, while it would be foolish not to employ the symbols for purposes of discovery, the view taken in the present volumes is, that the object of a geometrical theory is to reach such a comprehensive scheme of conceptions, that every geometrical result may be obvious on geometrical grounds alone.

The symbolical representation of the Propositions of Incidence. In the symbolical scheme which we now proceed to construct to accompany the geometry, we represent a point by a symbol, generally indicated by the same capital letter as that by which we name the point. But we suppose that, if P is the symbol which represents a point, precisely the same point is equally represented by a symbol mP, where m is any one, other than 0, of the system of algebraical symbols which we are employing. It is supposed that this system has no singular symbol other than 0, so that m^{-1} is also a symbol of the system. If P, Q represent any two points, we regard the symbol $mP + nQ$, where m and n are any two algebraical symbols of the system employed, as equally representing a point; and regard all points so represented, for different values of m and n, as being points of the line determined by the points P and Q, making the further assumption that all points of this line can be so represented. Thus a symbol $mP + nQ$ can be replaced by a symbol $-kR$, where R represents a point of the line PQ and k is an algebraic symbol of our system; and, conversely, if P, Q, R be points of a line, there is a relation between them which we may express by writing

$$mP + nQ + kR = 0,$$

speaking of this expression as a *syzygy*. Here m, n, k are algebraic symbols of the system employed; in general, when such a syzygy arises, it is in the course of work in which a precise symbol, not further susceptible of being multiplied by an algebraic symbol, has been introduced for each of the points P, Q, R; when this is so, the symbols m, n, k in the syzygy are definite when the points P, Q, R are given, save for a common multiplier prefixed to each. Such syzygies we suppose capable of being combined together, by addition and subtraction, in the manner of linear algebraic equations, the order of the terms in any syzygy being indifferent. To this end it is necessary to understand that a symbol $mP + nP$ may be replaced by $(m + n) P$, and, more generally, that a symbol $m\,(nP)$ may be replaced by $mn\,.\,P$. This last assumption requires the associative law in multiplication for the algebraic symbols; for, in virtue of this,

$$m\,[n\,(kP)] = m\,[nk\,.\,P] = m\,(nk)\,.\,P,$$

and
$$m\,[n\,(kP)] = mn\,(kP) = (mn)\,k\,.\,P.$$

Similarly the associative law in addition is involved in the former assumption. For $[(m + n) + k]\,P$ is thence $(m + n)\,P + kP$, and this is $mP + nP + kP$, which again is $[m + (n + k)]\,P$.

These various suppositions receive their justification by the exact correspondence they set up between the Propositions of Incidence and their symbolical representation.

In general when a point is, in the phraseology already employed in formulating the Propositions of Incidence, *dependent* upon other points, there is a syzygy connecting its symbol with those of these other points; and, conversely, points whose symbols are connected by a syzygy are not independent. For instance, three points which are not in line determine a plane, but for any four points of a plane, represented by symbols A, B, C, D, there is a syzygy

$$aA + bB + cC + dD = 0,$$

in which a, b, c, d are algebraic symbols of the system employed; and any point, D, of the plane determined by the points A, B, C, has a symbol capable of the form

$$D = -d^{-1}aA - d^{-1}bB - d^{-1}cC,$$
or, say, $\qquad D = mA + nB + kC.$

Such a statement may be expressed also by saying that the point D is a *derivative* of the points A, B, C.

The formulation is applicable to space of any number of dimensions. Any $r + 1$ independent points, having symbols unconnected by a syzygy, determine a space of r dimensions, in which any $r + 2$ points have symbols which are so connected, and conversely. If such $(r + 2)$ points be A_1, A_2, ..., A_{r+2}, this syzygy, written in the form

$$m_1A_1 + m_2A_2 = m_3A_3 + \ldots + m_{r+2}A_{r+2},$$

expresses that the line determined by the two points A_1, A_2, has a point in common with the $(r - 1)$ fold determined by the r points A_3, A_4, ..., A_{r+2}.

Examples of the application of the symbolism. Particular examples may be more illustrative than any further general statement:

(*a*) That two planes in a threefold space have a line in common may be formulated thus: let the planes be respectively determined by the points A, B, C, and the points A', B', C'. As, in three dimensions, any five points are connected by a syzygy, there is such a syzygy connecting the points A', B' and A, B, C; by this is determined a point of the line $A'B'$ which also lies on the plane ABC; this point is then on both planes. Another such point is determined by the syzygy connecting the five points B', C' and A, B, C. The line joining these two points is then that common to the two planes.

(*b*) That, in threefold space, a definite transversal can be drawn through an arbitrary point to meet two lines which have no point in common, is clear because, if the point be P, and the lines be, respectively, determined by the pairs of points A, B and C, D, the syzygy connecting these five points, of the form

$$pP + aA + bB + cC + dD = 0,$$

expresses that the three points P, $aA + bB$, $cC + dD$, being connected by a syzygy, are in line. The points $aA + bB$, $cC + dD$ are upon the two lines, respectively.

(c) Desargues' theorem, that, if two triads of points, A, B, C and A', B', C', be in perspective, the joins of corresponding points, of B, C and of B', C', of C, A and of C', A', of A, B and of A', B', meet in three points which are in line, also follows easily. For, if the lines AA', BB', CC' meet in O, the symbols for A', B', C' must be, respectively, of the forms

$$A' = \alpha A + \lambda O, \quad B' = \beta B + \mu O, \quad C' = \gamma C + \nu O,$$

where α, λ, β, μ, γ, ν are algebraic symbols. There will then, in accordance with what has been said, be no loss of generality in using A', B', C' instead of $\lambda^{-1}A'$, $\mu^{-1}B'$, $\nu^{-1}C'$, respectively, so that, if a, b, c be used for $\lambda^{-1}\alpha$, $\mu^{-1}\beta$, $\nu^{-1}\gamma$, respectively, these syzygies become

$$A' = aA + O, \quad B' = bB + O, \quad C' = cC + O.$$

These give however

$$B' - C' = bB - cC, \quad C' - A' = cC - aA, \quad A' - B' = aA - bB,$$

which, respectively, determine the three points of intersection of corresponding joins, and shew, in virtue of the syzygy

$$(B' - C') + (C' - A') + (A' - B') = 0,$$

that these three intersections are in line.

This example is instructive because the proof is applicable whether the two triads A, B, C and A', B', C' are in one plane or not; in the former case there is a syzygy connecting O with A, B, C, but the proof does not need to utilise this. It has been remarked however, and a formal proof will be given in Chapter II, that Desargues' theorem is incapable of proof from the Propositions of Incidence applicable to a plane only. The symbolism we are using thus implies more than these, even when applied only to a plane. Evidently it implies that any plane through two points contains every point of the line joining these points.

(d) In space of four dimensions two planes have a point in common; for if these planes be respectively determined by the points A, B, C and A', B', C', the two members of the syzygy which necessarily connects these six points in four dimensions,

$$aA + bB + cC = a'A' + b'B' + c'C',$$

represent points lying respectively, in these planes, which, by virtue of the syzygy, are identical.

Or again, if three lines in fourfold space be respectively given by the pairs of points P and P', Q and Q', R and R', the syzygy connecting these six points,

$$pP + p'P' + qQ + q'Q' + rR + r'R' = 0,$$

itself expresses that the three points $pP + p'P'$, $qQ + q'Q'$, $rR + r'R'$, lying respectively on these lines, are in syzygy. So that three lines of general position in fourfold space have a common transversal.

Interpretation of the fundamental laws of the algebraic symbols. After these illustrative examples, which could be indefinitely multiplied, we pass now to give a geometric interpretation of the laws of operation of the algebraic symbols which have been formulated.

(1) The symbol $-a$.

Let O, U be any two points, and P, represented by $O + aU$, be

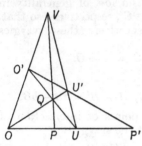

any point on the line joining these. As has been remarked, incidentally, in the above consideration of Desargues' theorem, there would be no gain in representing P in the form $mO + aU$. In any plane through the line OU take the points U', V, on a line passing through U; let OU', VP meet in Q, then QU and OV meet in O', and lastly $O'U'$ meet OU in P'. Thus P' is the harmonic conjugate of P in regard to O and U.

We may regard the points of the plane as dependent upon O, U and V. The symbol for the point U' is of the form $U + mV$, and by putting V for mV, in accordance with our original convention, we may suppose this symbol to be $U' = U + V$. Then the point Q is in syzygy with O and U', and also in syzygy with P and V, and the symbol for Q is determined by the fact that it expresses both these circumstances; absorbing an immaterial multiplier, we thence have $Q = O + a(U + V)$, which is both $O + aU'$ and $P + aV$. Similarly, the symbol for O', which is in syzygy both with O and V, and with U and Q, is given by $O' = O + aV$, which is the same as $Q - aU$. Thence the point P', which is in syzygy as well with O' and U', as with O and U, is given by $P' = O - aU$, this being the same as $O + aV - a(U + V)$, or $O' - aU'$.

Thus the point $O - aU$ is P'.

(2) The symbol $a + b$.

Let the points A, B of the line joining the points O and U,

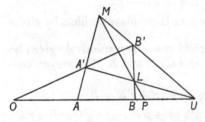

when referred to O and U, be, respectively, represented by symbols

$$A = O + aU, \quad B = O + bU.$$

We desire to construct the point $O + (a + b)U$.

In an arbitrary plane drawn through the line AB, draw two lines respectively through

A and B, and let any line through O, in this plane, meet these respectively in A' and B'. Let UA' meet BB' in L, and UB' meet AA' in M, and let LM meet the original line OU in P.

The figure is evidently determinate when, beside O, U, A, B, the points M and A' are given. For the symbol of M we may take $M = A + mA'$, or $M = O + aU + mA'$, which gives $M - aU = O + mA'$; in this syzygy the two members are the symbols, respectively, of a point on the line MU and of a point on the line OA'; this point must then be B', and we may write, for the symbol of this, $B' = O + mA'$. As, however, we have $B = O + bU$, we thence obtain $B' - B = mA' - bU$, which states the equivalence of a point of the line $B'B$ with a point of the line $A'U$. We can, therefore, write, for the symbol of the point L the form $L = mA' - bU$. Thus, as $M = A + mA'$, and, therefore, $M - L = A + bU$, we can, similarly, infer that the point P is given by $P = A + bU$, or $P = O + (a + b)U$.

The construction of P is clearly symmetrical in regard to the points A, B, in accordance with the equation $a + b = b + a$, and, as in earlier cases considered in Section I, is independent of the lines drawn through A, B, O, U when these points are given. Further, when b is $-a$, the figure at once gives $P = 0$, in accordance with $a + (-a) = 0$.

(3) The equation $(a + b) + c = a + (b + c)$.

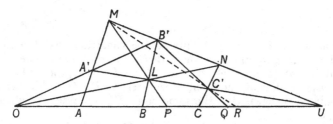

Let p, q denote $a + b$ and $b + c$, respectively. Let A, B, C, referred to the points O, U be, respectively,

$$A = O + aU, \quad B = O + bU, \quad C = O + cU.$$

Make the same construction as in the preceding case for A and B, so obtaining the point P given by $P = O + pU$. Now let OL meet the line $MB'U$ in N; join CN meeting the line $A'LU$ in C'; join $B'C'$ meeting OU in Q; also join MC', meeting OU in R. Then as the points L, N, on BB', CC', respectively, are in line with O, it follows, as in the preceding case, that Q is $O + qU$.

Further, as the points A', B', on the lines AM, QC', are in line with O, it follows, also as in the preceding case, that $R = O + (a + q)U$. And, as the points L, N, on the lines PM, CC', are in line with O, it follows, as in the preceding case, that $R = O + (p + c)\,U$. Thus $p + c = a + q$, as was desired.

(4) The interpretation of the product ab.

For the interpretation of the product ab we go back to a figure which we have already been twice concerned with, first as an immediate application of the Propositions of Incidence (p. 16 above), and then as an example of the consequences of Pappus' theorem (p. 60 above).

Upon the line OU let the points A, B be respectively $O + aU$ and $O + bB$, and let E be the point $O + U$. In an arbitrary plane through this line, draw lines o, u, a, b, respectively through the

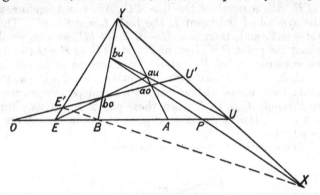

points O, U, A, B, subject only to the condition that the intersections (a, u), (b, o) are in line with E. Let the joining line of the points (a, o), (b, u) meet the line OU in P. We proceed to shew that the symbol of the point P is given by $P = O + abU$. It has already been shewn that the same point P is obtained for all lines o, u, a, b.

Let the lines a, b meet in the point Y, and the line o meet YU in the point U', and meet the line YE in the point E'. The two triads of points E', E, B and (a, o), (a, u), (b, u) are in perspective from the point (b, o); and the corresponding joins, of E' to E and (a, o) to (a, u), meet in Y, while the corresponding joins, of B to E and (b, u) to (a, u), meet in U. Wherefore, by Desargues' theorem, the line $E'B$ meets the join of (a, o) to (b, u) in a point of the line UY, say X.

The symbol of U', by absorption of a multiplier in the symbol of Y, may be taken to be $U' = U + Y$. Then E', being both on the join of O to $U + Y$, and on the join of E, or $O + U$, to Y, is given by $E' = O + U + Y$. In a similar way, the point (a, o), being on the join of O to U', or $U + Y$, and also on the join of A, or $O + aU$, to Y, has a symbol $O + aU'$, which is $O + aU + aY$. We find, likewise, for the symbol of the point (b, o), the form $O + bU'$. Thence, the point X, being on the line $U'U$, and also on the join of B to E', that is of $O + bU$ to $O + U'$, has the symbol $U' - bU$. Thence, finally, the point P, being on the line OU, and also on the join of X to (a, o), that is of $U' - bU$ to $O + aU'$, has a symbol

$$O + aU' - a(U' - bU), \quad \text{or} \quad O + abU.$$

If a line, o', be drawn through O to meet the line a in a point (a, o') which is in line with E and (b, u), and the point (b, o') in which this meets the line b, be joined to the point (a, u) by a line meeting OU in the point P', it can be similarly shewn that the line $E'A$ meets the join of (a, u) and (b, o') in a point of the line YU, and that the point P' has a symbol $O + baU$.

(5) The possibility of the equation $ab = ba$.

We have seen above that the points P, P' coincide if we assume Pappus' theorem (pp. 60, 61). In the present description of the derivation of the points P and P', we have slightly added to the original figure of p. 16, using Desargues' theorem, in order to simplify the proof of the symbols for the points P, P'. In the figure thus reached, arbitrary lines are drawn from E, B, A, U to

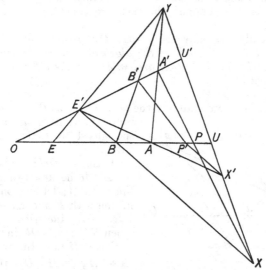

meet in Y; an arbitrary line is drawn from O to meet these, respectively, in E', B', A', U'; then P is found by drawing $E'B$ to meet YU in X, and taking the intersection of $A'X$ with OU, while P' is found by drawing $E'A$ to meet YU in X', and taking the intersection of $B'X'$ with OU. The identity of P and P' as a consequence of Pappus' theorem can then be seen, differently from

before, by considering the two triads Y, X, X' and E', B', A'; as
$YA', E'X'$ meet in A, while $YB', E'X$ meet in B, it follows then
that XA' and $X'B'$, or $A'P$ and $B'P'$, meet on AB.

A direct deduction from the figure of p. 16 of the symbols for
P and P' may be given, which has a value as an exhibition of the
self-consistency of the non-commutative symbolism.

A further construction of P is obtainable by remarking that, in
virtue of Desargues' theorem, the lines UE', UA' meet YB, YP,
respectively, in two points lying in line with O.

We have seen above (p. 46) that Pappus' theorem, though a
theorem of geometry in a plane, is equivalent with a theorem in
regard to lines in space. We shall see below (p. 82) that the corresponding expression in symbols, of this theorem in regard to lines
in space, also requires that the symbols should be commutative in
multiplication.

We have remarked above (p. 66) that, in the case of the iterative
symbols, the commutative equation $mn = nm$ is a consequence of
their definition. The question whether the geometrical representation of this identity involves Pappus' theorem, and the question
whether von Staudt's introduction of numbers in Geometry involves
the assumption of Pappus' theorem, are considered below (p. 87
and p. 93).

Provisionally, we may infer that *the assumption of Pappus'
theorem, and the restriction of the algebraic symbols to such as are
commutative in multiplication are, subject to the other assumptions
made, equivalent.*

(6) The symbol a^{-1}.

This may be represented by the same figure as that used for the
representation of ab, if this be
specialised so that P coincides
with E.

On the line OU let A be $O+aU$;
let A', H be any two points in
line with A; let B', on OA', be
in line with E and H, and N, on
UH, be in line with E and A'.
Then NB' meets OU in $O + a^{-1}U$.

For, if H have the symbol

$$A + \lambda A', \text{ or } H = O + aU + \lambda A',$$

we first have

$$(1 - a) O + \lambda A' = O + aU + \lambda A' - a (O + U),$$

as the symbol of a point lying both on OA' and on HE, the point
B'; and then we have

$$O + U + \lambda A' = O + aU + \lambda A' + (1 - a) U$$

as the symbol of a point lying on EA' and also on HU, the point N. Thence we have $aO + U$ as the symbol of a point lying on $B'N$ which lies on OU; and this gives the same point, A_1, as $O + a^{-1}U$.

Evidently, by the construction, if $b = a^{-1}$, then $a = b^{-1}$, and $ab = 1$. Also when $a = 1$, then also $a^{-1} = 1$.

(7) The equation $(ab)c = a(bc)$.

Let p denote ab, and q denote bc.

Let the points A, B, C and E, on the line OU be respectively $O + aU, O + bU, O + cU$, and $O + U$. Draw, in an arbitrary plane

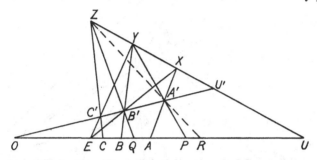

through OU, two lines through O and U, intersecting in U', and let a line through B, in this plane, intersect these, respectively, in B' and Y. Let the lines joining the point E, respectively to Y and B', meet these lines, OU', UU', respectively in C' and X. Let AX meet OU' in A', and CC' meet UU' in Z. Lastly, draw YA', ZB', ZA', meeting OU respectively in P, Q, R.

Then, recalling once more the construction for $O + mnU$ from the points $O + mU, O + nU,$ we see (a) that P is $O + abU$, or $O + pU$; (b) that Q is $O + bcU$, or $O + qU$; (c) that R is $O + aqU$; (d) that R is also $O + pcU$. Wherefore $aq = pc$, as we desired to see.

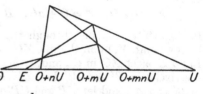

(8) The equation $(a + b)c = ac + bc$.

First remark the following figure: let $ZPR, ZR'P', UPR', URP'$ be any four lines in a plane, with intersections indicated by the lettering, the joins PP', RR' meeting in T; take an arbitrary point, Y, upon ZU, and an arbitrary point, C', upon PP'; let the joins YP, YP' meet $P'R, R'P$ respectively in the points Q and Q', so that, by Desargues' theorem, as corresponding joins of the two triads P, Q, R and P', Q', R' meet, respectively, in the three points U, Z, Y, which are in line, it follows that QQ' passes through the point of intersection, T, of PP' and RR'; next, let the joins $C'Y, C'Z$ meet QQ'

and RR', respectively, in H and K. Then, as the corresponding joins of the two triads Y, H, Q' and Z, K, R' meet, respectively,

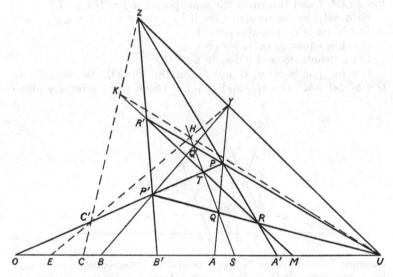

in the points T, P', C', which are in line, it follows that the line KH passes through the intersection, U, of ZY and $R'Q'$. We are concerned now with the intersections of all the lines of this figure with a line drawn through U, in the plane of the figure.

Let O, U be any two points, and E, A, B, C be points upon the line OU, whose symbols, referred to O and U, are respectively

$$E = O + U, \quad A = O + aU, \quad B = O + bU, \quad C = O + cU.$$

In an arbitrary plane through the line OU take a point, Z, and then a point, Y, upon the line ZU. Join YB and YA, and also ZC. Let YE meet ZC in C', and, then, let OC' meet YB and YA, respectively, in P' and P. Let ZP' and ZP, respectively, meet OU in B' and A'; and let UP and UP' meet ZP' and ZP, respectively, in R' and R, these same lines meeting YP' and YP, respectively, in Q' and Q. Then, by what has been remarked, the line QQ' passes through the intersection, T, of PP' and RR', and if this line QQ' meet $YC'E$ in H, while RR' meets $ZC'C$ in K, then KH passes through U. Denote the intersections of QQ' and RR', respectively, with OU by S and M.

We have just recalled (in No. 7) the construction for finding the point $O + mnU$ from the points $O + mU$ and $O + nU$; if we recall also the construction for finding $O + (m + n)U$ from $O + mU$ and $O + nU$, we see that, in the figure we have constructed, the symbol of the point S is $O + (a + b) U$, and, thence, that the symbol of the point M is $O + (a + b) cU$. On the other hand, we see also that the symbol of the point A' is $O + acU$, and, similarly, of B' is $O + bcU$, and, thence, that the symbol of the point M is

O $O+nU$ $O+mU$ $O+(m+n)U$ U

$O + (ac + bc) U$. Comparing these two forms for the symbol of the point M, we have the identity $(a + b) c = ac + bc$, which we desired to interpret.

A similar representation of the identity $c (a + b) = ca + cb$ may be constructed.

Representation of the algebraic effect of Pappus' Theorem in three dimensions. We have now shewn that all the fundamental laws of operation for the algebraic symbols are in accord with geometrical facts in one plane deducible from the Propositions of Incidence with the help of Desargues' theorem, which is itself deducible, as we have seen, from these Propositions, assumed not limited to two dimensions. But we have shewn that the commutative law of multiplication for the algebraic symbols involves Pappus' theorem, and conversely. We shewed however, in Section II, that Pappus' theorem, is equivalent, in virtue of the Propositions of Incidence, to a theorem of intersection of two lines in three dimensions. It is interesting to verify directly that this theorem again involves the commutative property of the symbols. The geometrical theorem is that, if we have two triads, each of non-intersecting lines, in three dimensions, every line of either triad meeting all those of the other, then any transversal of the lines of one triad intersects any transversal of the lines of the other.

Let the lines of the two triads be respectively a, b, c and a', b', c', their intersections being, A, A', A'' on a, B, B', B'' on b, and C, C', C'' on c. Let x be any transversal of a, b, c, meeting them, respectively, in P, Q, R, any point of this transversal being X; so, let y be any transversal of a', b', c', meeting them, respectively, in P', Q', R', any point of this transversal being Y.

The points A, A', B, B' are not supposed to be in a plane, and may be taken as fundamental for the threefold space in which the figure lies. Using A, A', B, B' as the symbols for these points, we may, by absorption of algebraic multiples in these symbols, suppose that the symbol of the point C'' is

$$C'' = A + A' + B + B'.$$

This expresses, however, that C'' is a derivative of the two points represented, respectively, by $A + A'$ and $B + B'$, of which the former is on the line a and the latter on the line b. For the symbols of the points A'' and B'' we thus infer

$$A'' = A + A', \quad B'' = B + B'.$$

For the symbols of the points C and C' we similarly infer

$$C = A + B, \quad C' = A' + B'.$$

The points P and Q are, respectively, derivatives of A, A' and of B, B'; thus proper symbols for these are, respectively,

$$P = A + \rho A', \quad Q = B + \lambda B',$$

with suitable values for the algebraic symbols ρ and λ. Then the point R, a derivative of P and Q, must have a symbol of the form

$$R = mP + nQ, \quad = mA + nB + m\rho A' + n\lambda B';$$

on the other hand, as R is a derivative of C and C', its symbol must be of the form

$$R = h(A + A') + h'(B + B');$$

comparing these two forms, we infer that $m = n$ and $m\rho = n\lambda$, and hence $\rho = \lambda$. Thus the symbols of P, Q involve the same algebraic symbol, being $P = A + \rho A'$, $Q = B + \rho B'$.

Thence the point X, a derivative of P and Q, is given by a symbol

$$X = P + \sigma Q, \quad = A + \rho A' + \sigma B + \sigma \rho B'.$$

Again, the points P' and Q', respectively derivatives of A, B and of A', B', have symbols of the forms

$$P' = A + \kappa B \quad \text{and} \quad Q' = A' + \mu B';$$

and the point R', a derivative of P' and Q', is equally a derivative of A'' and B''. There exists, therefore, a syzygy of the form

$$p(A + \kappa B) + q(A' + \mu B') = r(A + A') + s(B + B');$$

hence we infer that $p = q$ and $p\kappa = q\mu$, and therefore that $\kappa = \mu$.

Thus the point Y, any point of the line $P'Q'$, has a symbol

$$Y = P' + \tau Q', \quad = A + \tau A' + \kappa B + \tau \kappa B'.$$

Comparing the symbols for the points X and Y, we see that the conditions for the lines x and y to have a common point are that, when ρ and κ have been taken, corresponding to the positions of P and P' respectively on AA' and AB, it must be possible to take σ and τ, corresponding to the positions of X and Y respectively on x and y, so that

$$\tau = \rho, \quad \kappa = \sigma, \quad \tau\kappa = \sigma\rho.$$

These can be satisfied if, and only if, we have

$$\rho\kappa = \kappa\rho,$$

and therefore, $\sigma\tau = \tau\sigma$, $\rho\sigma = \sigma\rho$, $\kappa\tau = \tau\kappa$. The point X is determined by ρ and σ alone; only when these are such that $\rho\sigma = \sigma\rho$ is there a line y, transversal to a', b', c', which meets the line x. The condition of commutative multiplication, for every two of the algebraic symbols employed, is the necessary and sufficient condition that every line x should intersect every line y.

Examples in regard to the use of the symbols, in particular in regard to the commutative law of multiplication. A vast number of applications can be given of the symbolism explained in this Section. We limit ourselves mainly to such as seem to carry the explanation of general principles somewhat further.

Ex. 1. In the construction above given for the point $O + (a+b)U$,

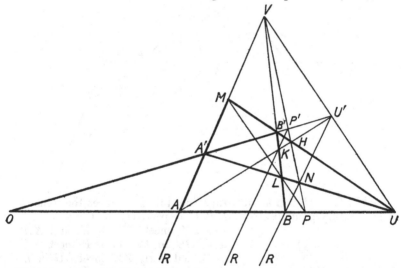

from the points O, U, $O + aU$ and $O + bU$, the constructed point P may be differently defined. If AA', BB' meet in V, also $A'B'$ meets VU in U', and meets VP in P', while $U'A$, $U'B$ meet VP, respectively, in H and N, and $U'A$ meets VB in K, then H and N lie, respectively, on UB' and UA', and the lines $P'K$, NB, VA meet in one point, R.

These follow by applying Desargues' theorem.

Ex. 2. The construction of the

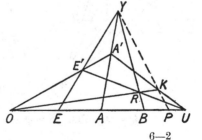

6—2

point $O + abU$, from the points $O, U, E, O + aU, O + bU$, above given, may be stated thus: Let $OE'A'$ be any line through O meeting YE, YA, respectively, in E' and A', where Y itself is arbitrary; let UE' meet YB in R, and OR meet UA' in K; then YK gives the required point P. To reduce this to what is given above we apply Desargues' theorem to the two triads E', R, B and A', K, P.

Ex. 3. If O, U, E, A, B, C represent $O, U, O + U, O + aU$,

$O + bU, O + cU$, and an arbitrary line through O meet the joins of the arbitrary point, Y, to E, A, B, C, respectively, in E', A', B', C', also $E'C$ meet YU in X, and XA' meet OU in P, and $B'P$ meet YU in Z, and lastly ZE' meet OU in L, shew that the symbol of this point is $O + b^{-1}acU$. Then construct a geometrical representation of the equation

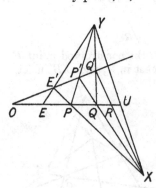

$$b^{-1}ac + b^{-1}dc = b^{-1}(a + d)c.$$

Ex. 4. If relatively to the points O, U, the points

$$E = O + U, \quad P = O + aU$$

be given, and $a^2, a^3, \ldots$ as usual denote $a.a, a^2.a, \ldots$, construct the points

$$Q = O + a^2U, \quad R = O + a^3U, \quad S = O + a^4U, \text{ etc.}$$

Ex. 5. To represent the equations

$$(-a)(-b) = ab,$$
$$(-a)b = a(-b) = -ab.$$

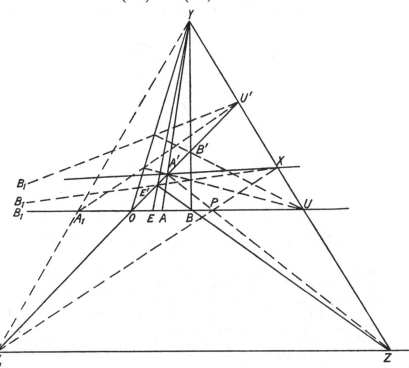

Let $\qquad E = O + U, \quad A = O + aU, \quad B = O + bU.$

Let Y be any point, not on OU, and U' be any point on YU. Let the line OU' meet YE, YA, YB, respectively, in E', A', B'; let $E'B$ meet YU in Z, and $A'Z$ meet OU in P. Join the intersection of YO and UA' to U', the join meeting OU in A_1; join the intersection of YO and UB' to U', the join meeting OU in B_1. Let YA_1 meet OU' in A_1'; let B_1E' meet YU in X.

Prove that $A_1'X$ passes through P, and hence represent the equation $(-a)(-b) = ab$. Also prove that $A'X$ and $A_1'Z$ meet on OU, and hence represent the equation $(-a)b = a(-b)$.

Ex. 6. The iterative symbols.

(*a*) Let m, n be any iterative symbols formed, respectively, by $\overline{m}$ and by $\overline{n}$ repetitions of the symbol 1, where $\overline{m}$, $\overline{n}$ are positive integers of ordinary arithmetic. In the first place, given the points O, U

and $E = O + U$, we seek to represent the points $O + mU$, $mO + U$, $mO + nU$.

Take any point Y, not on OU, and any point T on YE, for which the symbol, with proper choice of the symbol Y, may be written $T = Y + E$. Let OT meet YU in H, and UT meet OY in K; the

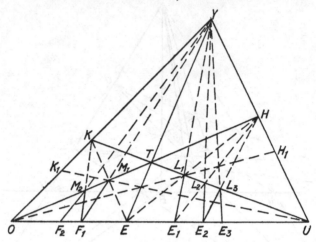

point H, on the join of U and Y, and on the join of O and $Y+O+U$, has symbol $H = U + Y$; so the point K has symbol $K = O + Y$. Let HE and KE meet UT and OT, respectively, in L_1 and M_1; and let YL_1 and YM_1 meet OU in E_1 and F_1, respectively. Let HE_1 and KF_1 meet UT and OT, respectively, in L_2 and M_2; and let YL_2 and YM_2 meet OU in E_2 and F_2, respectively. And let the process thus suggested be continued indefinitely.

It can then be seen at once that the symbols for E_1, E_2, ... are $E_1 = O + 2U$, $E_2 = O + 3U$, ..., in general $E_{r-1} = O + rU$. So we have $F_1 = 2O + U$, $F_2 = 3O + U$, ... Then, if the join of O and L_1 meet YU in H_1, the join of K to H_1 meets OU in $O - 2U$; and, in general, if the join of O and L_{r-1} meet YU in H_{r-1}, the join of K to H_{r-1} meets OU in $O - rU$. Similarly if the join of U and M_1 meet YO in K_1, the join of H to K_1 meets OU in $2O - U$, and so on. We thus easily construct all the points $O + mU$, $mO + U$, $O - mU$, $mO - U$.

In fact these points arise by successive applications of the construction for determining the harmonic conjugate of a point P in regard to a given pair of points A, B. For, denoting this fourth harmonic point by $(A, B)/P$, we clearly have

$$E_1 = (E, U)/O, \quad E_2 = (E_1, U)/E, ...,$$
$$F_1 = (E, O)/U, \quad F_2 = (F_1, O)/E, ...,$$

and in general

$$E_{r+1} = (E_r, U)/E_{r-1}, \quad F_{r+1} = (F_r, O)/F_{r-1}.$$

More generally, recalling that the harmonic conjugate of a point $A + \lambda B$, in regard to the points A, B, is $A - \lambda B$, we see that the point $mO + nU$, being $mO + (n-1)U + U$, is the harmonic conjugate of $mO + (n-2)U$ in regard to $mO + (n-1)U$ and U, or, say,

$$E_{m,n} = (E_{m, n-1}, U)/E_{m, n-2}.$$

Every one of the points $mO \pm nU$ can thus be constructed by a succession of processes of finding the fourth harmonic, given the points O, U, $O + U$; the m, n being any iterative symbols.

Consider next the representation of the equation $nm = mn$; we shew that this is independent of Pappus' theorem. In the first place, from the points O, U, E, and Y, we can construct, as above, the points $mU + Y$, $nU + Y$, upon the line YU, and the points $O + nY$, $O + mY$, upon the line YO. The intersection, P, of the joins of O to $mU + Y$ and of U to $O + nY$, evidently has a symbol

$$P = O + nmU + nY,$$

and the intersection, Q, of the joins of O to $nU + Y$ and of U to $O + mY$, evidently has a symbol

$$Q = O + mnU + mY.$$

A representation of $nm = mn$ is, thus, that the points P, Q are in line with Y.

If, however, we consider the three points, P, so arising, for the same value of n, and the three values $m - 1$, m, $m + 1$ in place of m, and at the same time the three points, Q, so arising, for this value of n and the same values for m, the harmonic relation shews that the assumption that two of the three triads Y, P, Q so arising are in line, involves that the third is also. For the initial values of m, n the theorem that Y, P, Q are in line is easy to prove. Thus, by induction, it is always true.

Ex. 7. As examples of the fact that a geometrical result obtainable without Pappus' theorem should be representable symbolically without use of the commutative law for multiplication, and the converse statement that, if we can obtain a symbolical deduction of a result without use of this commutative law, we should be able to construct a geometrical proof from the Propositions of Incidence only, we may reconsider some of the work of Section I.

A simple result was that, if p, a, b, c, d be five arbitrary lines of which no two intersect, lying in three dimensions, and, from an arbitrary point, P, of the first line, p, there be drawn the transversal to two of these lines, say a and c, to meet these respectively in A and C, and also from P the transversal to the other two given lines b, d meeting these in B and D, then the intersection, say Q, of the joins of A and C respectively to B and D, that is of AB and CD, also describes a line. The intersection of AD and CB equally describes a line.

Choosing the symbols of A, B, C, D suitably, we can suppose, these points being in a plane, that the syzygy connecting them is $A + B + C + D = 0$. Then the point represented equally by $A + C$ and $-(B + D)$ is necessarily P; and, similarly, for the symbol of Q, we have either $A + B$ or $-(C + D)$. If a second point, P', be taken on the line p, and a similar construction, with a like convention, be made, we can equally write

$$P' = A' + C', \; = -(B' + D'), \quad Q' = A' + B', \; = -(C' + D').$$

A third point, P'', upon p, has a symbol given by

$$P'' = mP + m'P' = m(A + C) + m'(A' + C')$$
$$= mA + m'A' + mC + m'C',$$

which is also $\quad -(mB + m'B') - (mD + m'D')$.

Thus P'' is in line with the points $mA + m'A'$ and $mC + m'C'$, lying, respectively, on the lines a and c, and is also in line with the points $mB + m'B'$ and $mD + m'D'$, lying, respectively, on the lines b and d. These are then the points obtainable by the original construction from the point P''. If we denote them, respectively, by A'', C'', B'', D'', we see that

$$A'' + B'' = m(A + B) + m'(A' + B') = -(C'' + D''),$$

and the lines $A''B''$, $C''D''$ meet in a point, Q'', given by

$$Q'' = mQ + m'Q',$$

lying on the join of Q and Q'. This establishes the result stated.

Ex. 8. In Section i (p. 27) we gave a construction, when four, or more, non-intersecting lines AA', BB', CC', DD' ..., have three common transversals $ABCD$ ..., $A'B'C'D'$ ..., $PQRS$ ..., for finding a fourth transversal $P'Q'R'S'$ .... The point P' was the harmonic conjugate of P in regard to A and A', say $P' = (A, A')/P$; and, similarly, $Q' = (B, B')/Q$, $R' = (C, C')/R$, ...

The figure being in three dimensions we can regard A, A', B, B' as fundamental points, and write the symbols of P and Q in the forms $P = A + \rho A'$ and $Q = B + \sigma B'$; and then the symbols of R and S in the forms $R = P + mQ$ and $S = P + nQ$. By properly choosing the symbols A', B', Q, we can, if we wish, suppose $\rho = 1$,

$\sigma = 1$, $m = 1$; but n is not 1, save for a particular position of S. The form of R,

$$R = A + \rho A' + m(B + \sigma B'),$$

enables us then to infer, for the symbols of C and C', respectively,

$$C = A + mB, \quad C' = \rho A' + m\sigma B';$$

also if, as before (p. 27 above), the transversal from R to the cross lines AB' and $A'B$ meet these respectively in F, F', this form of R shews that, for F and F',

$$F = A + m\sigma B', \quad F' = \rho A' + mB.$$

Hence

$$C - F' = A - \rho A', \; = F - C', \quad \text{and} \quad C - F = m(B - \sigma B'), \; = F' - C';$$

thus the lines $F'C$ and FC' meet in a point P' of AA' given by $P' = A - \rho A'$, and the lines FC and $F'C'$ meet in a point Q' of BB' given by $Q' = B - \sigma B'$. These are such that $P' = (A, A')/P$ and $Q' = (B, B')/Q$; and they give

$$P' + mQ' = C - mC',$$

which shews that $P'Q'$ meets CC' in a point R' such that $R' = (C, C')/R$. From $S = P + nQ$, we can, similarly, shew that the line $P'Q'$ contains a point S' for which $S' = (D, D')/S$.

Ex. 9. Another construction was given (p. 29 above) for a fourth transversal of the four lines AA', BB', CC', DD' of Ex. 8. If BR, $B'S$ meet in X, and XP meets BB' in Q'; while $A'R$, AS meet in Y, and YQ meets AA' in P', then $P'Q'$, which meets AA' and BB', also meets CC' and DD'. Further XA, YB', CC', $P'Q'$ meet in a point, as do YB, XA', DD', $P'Q'$.

With the notation of the previous Example we find, in fact,

$$X = R - (m-n)B = S + (m-n)\sigma B' = A + \rho A' + nB + m\sigma B', \; = D + C',$$

so that

$$nQ' = nB + m\sigma B', \quad Q' = B + n^{-1}m\sigma B'.$$

Also we find

$$Y = m^{-1}R - (m^{-1} - n^{-1})\rho A' = n^{-1}S + (m^{-1} - n^{-1})A$$
$$= m^{-1}A + n^{-1}\rho A' + B + \sigma B', \; = m^{-1}C + n^{-1}D',$$

so that

$$m^{-1}P' = m^{-1}A + n^{-1}\rho A', \quad P' = A + mn^{-1}\rho A'.$$

These lead to

$$P' + mQ' = C + mn^{-1}C', \quad P' + nQ' = D + mn^{-1}D',$$

which we may denote, respectively, by R' and S'. These express that $P'Q'$ meets CC' and DD'.

But, unless we assume a relation, in multiplication, for the symbols, which would follow if we assumed Pappus' theorem, this line $P'Q'$ does not meet any other general transversal of the three given lines $ABCD$, $A'B'C'D'$, $PQRS$. For if, from a point, T, of the line $PQRS$, there be drawn ETE' to meet $ABCD$ and $A'B'C'D'$, respectively, in E and E', taking the symbol of T to be $T = P + tQ$, we obtain

$$E = A + tB, \quad E' = \rho A' + t\sigma B',$$

and $\quad P' + tQ' = E + mn^{-1}E' + mn^{-1}(nm^{-1}tn^{-1}m - t)\,\sigma B'.$

This shews that the point in which the line $P'Q'$ meets the plane $EE'B'$ is not on EE', that is, that $P'Q'$ does not intersect EE', unless

$$tn^{-1}m = mn^{-1}t,$$

a relation which, if we suppose $m = 1$, as we may do without loss of generality, becomes $tn^{-1} = n^{-1}t$, or $nt = tn$. It would follow if Pappus' theorem were assumed, and it is true, independently of this, when t, m, n are iterative symbols.

The other results stated to be true follow from the syzygies

$$X + (nm^{-1} - 1)\,A = nY + (m - n)\,\sigma B' = nm^{-1}C + C' = nQ' + nm^{-1}P',$$

these combinations being all identical with

$$nm^{-1}A + \rho A' + nB + m\sigma B',$$

together with

$$mY - (m - n)\,B = X + (mn^{-1} - 1)\,\rho A' = D + mn^{-1}D' = nQ' + P',$$

these being all equivalent with

$$A + mn^{-1}\rho A' + nB + m\sigma B'.$$

The facts that X lies on $C'D$, and Y lies on CD', have already appeared.

If we proceed by the same rule to deduce P'', Q'' from P', Q', just as we have deduced P', Q' from P, Q, we find that

$$P'' = A + mn^{-1}mn^{-1}\rho A', \quad Q'' = B + n^{-1}m\,n^{-1}m\sigma B',$$
$$P'' + mQ'' = C + mn^{-1}mn^{-1}C', \quad P'' + nQ'' = D + mn^{-1}mn^{-1}D',$$

the last two giving the points where $P''Q''$ meets CC' and DD'.

Ex. 10. We gave in Section I (p. 26) a process for finding a transversal, in the figure of Ex. 8, for the two pairs of cross joins AB', $A'B$, CD', $C'D$. With the notation of Ex. 8, we find for the symbols of the points M, M', N, N', occurring in that construction

$$M = A + m\sigma B', \quad M' = \rho A' + mB, \quad N = A + n\sigma B', \quad N' = \rho A' + nB,$$

from which follow

$$M \ + N' = C' + D, \quad M \ + mn^{-1}N' = C \ + mn^{-1}D',$$
$$M' + N \ = C + D', \quad M' + mn^{-1}N \ = C' + mn^{-1}D,$$

in agreement with the fact that the lines MN' and $M'N$ both meet the lines CD' and $C'D$.

Ex. 11. Consider now the question of two Moebius tetrads of points, A, B, C, D and A', B', C', D', such that each of the latter is on the plane determined by three of the former, and each of the former on the corresponding plane determined by three of the latter. We have, in Section I (p. 61), deduced the possibility of this by help of Pappus' theorem; it is to be expected that the symbolical representation will require the commutative restriction.

The points A, B, C, D being regarded as fundamental, the fact that the points D', A', B', C' are, respectively, on the planes ABC, BCD, CAD, ABD, can be expressed, by properly choosing the symbols of A, B, C, A', B', C', by the syzygies

$$D' = A + B + C,$$
$$A' = mC - n_1 B - D, \quad B' = nA - l_1 C - D, \quad C' = lB - m_1 A - D,$$

as is easily seen, l, m, n and l_1, m_1, n_1 being suitable algebraic symbols. Then, the line $B'C'$ meets the plane ABC in a point which is represented by either of the symbols

$$B' - C', \quad (m_1 + n) A - lB - l_1 C;$$

if this point is to be on the line $D'A$, remembering $D' = A + B + C$, this point must be a derivative of A and $B + C$. Thus $l = l_1$. So we get $m = m_1$ and $n = n_1$.

In order that D should lie on the plane $A'B'C'$, there must be a derivative of A', B', C', say $aA' + bB' + cC'$, which coincides with D. As A, B, C, D are not in a plane, there must then be a syzygy

$$a (mC - nB) + b (nA - lC) + c (lB - mA) = 0,$$

which, since A, B, C are independent, requires the equations

$$bn = cm, \quad cl = an, \quad am = bl;$$

these however require

$$a = cln^{-1}, \quad a = blm^{-1}, \quad cm . m^{-1} ln^{-1} = bn . n^{-1} lm^{-1},$$

and hence $m^{-1} ln^{-1} = n^{-1} lm^{-1}$, or $lm^{-1}n = nm^{-1}l$. By a slight change of symbols we can, without loss of generality, suppose $m = 1$. The condition for the Moebius tetrads is thus $ln = nl$, as was anticipated.

Ex. 12. Obtain the symbolical representation of the result given in Section I (p. 37) that, in four dimensions, an infinite number of planes can be drawn through an arbitrary point to meet three given lines of general position, and that these planes all meet three other lines.

Also of the result that, in five dimensions, an infinite number of planes can be drawn to meet four given lines of general position, and that these planes all meet six other lines.

Ex. 13. It has been seen that a plane meeting four lines, of general position, in four dimensions, can be drawn to meet two of these lines in assigned points (Section I, p. 37). Denote the four lines by a, b, c, d. Let the transversal of b, c, d meet b, c and d, respectively, in C, B' and P'; the transversal of c, a, d meet c, a, d respectively, in A, C' and Q'; and the transversal of a, b, d meet

a, b, d, respectively, in B, A' and R'. The points P', Q', R' are, thence, respectively, derivatives of B' and C, of C' and A, and of A' and B; they are however in syzygy among themselves, as being on the line d. This syzygy leads then to a syzygy connecting the symbols of A, A', B, B', C, C'; by a proper choice of the symbols this last syzygy may be taken to be $A + A' + B + B' + C + C' = 0$, any five of these points remaining independent, and sufficient to determine the fourfold space. Hence the symbol of P', a derivative of B' and C, may be taken to be $P' = B' + C$, with, similarly, $Q' = C' + A$, and $R' = A' + B$. The line a is determined by the points B and C', the line b by the points C and A', and the line c by the points A and B'.

If then we take the plane containing the three points

$$B + \rho C', \quad C + \sigma A', \quad A + \rho^{-1}\sigma^{-1}B',$$

we see, because

$$B + \rho C' + \sigma^{-1}(C + \sigma A') + \rho(A + \rho^{-1}\sigma^{-1}B')$$

is the same as

$$\sigma^{-1}(B' + C) + \rho(C' + A) + A' + B,$$

that this plane meets the line d. As ρ, σ may be taken arbitrarily we thus have the general plane meeting the four lines a, b, c and d, spoken of at starting.

With the four given lines is associated another line; it is that, namely, which contains the three points whose symbols are $A + A'$, $B + B'$, $C + C'$, these being in syzygy in virtue of the syzygy connecting the six points A, A', B, B', C, C'. This line may be defined geometrically; if a' be the common transversal of b, c, d; b' of c, a, d; c' of a, b, d and d' of a, b, c, it is the intersection of the four threefolds $\{a, a'\}$, $\{b, b'\}$, $\{c, c'\}$, $\{d, d'\}$, which meet in a line. For brevity we omit the geometrical proof of this, which will arise later; it is independent of Pappus' theorem.

But now we remark that

$$\rho^{-1}(B + \rho C') + C + \sigma A' + \sigma(A + \rho^{-1}\sigma^{-1}B')$$

is the same as

$$\sigma(A + A') + \rho^{-1}B + \sigma\rho\sigma^{-1}B' + C + C'.$$

The point of which this is the symbol will lie on the line containing the three points $A + A'$, $B + B'$, $C + C'$, only if $\rho^{-1} = \sigma\rho^{-1}\sigma^{-1}$, or $\rho\sigma = \sigma\rho$. Hence we forecast a geometrical theorem: *If, and only if, Pappus' theorem be assumed, all planes meeting four lines of general position in four dimensions are met by another line.*

A proof of this theorem, apart from the symbolism, will be given later.

Ex. 14. Finally we make a brief remark in regard to von Staudt's introduction of numbers in Geometry. This depends (*Beiträge zur*

Geometrie der Lage, Zweites Heft, No. 256, p. 166) upon considering two sets of four points, on two different lines, each set taken in a definite order, as being equivalent, when they are in perspective with one another. In effect such a set of four points, (or *Wurf*), is regarded as determining what we have called an algebraical symbol, and rules for the sum and product of two such symbols, or, in general, for computation with these symbols, are developed from geometrical theorems previously established on geometrical grounds.

It is impossible in a few words to make a comparison, between two methods of arranging the fundamental logical ideas, which shall not be open to objection. But it would seem that von Staudt's scheme, when approached from the point of view we have taken up, assumes Pappus' theorem in its initial definition. For, as will appear more fully below, for four points of a line which have, with the suppositions here made, the respective symbols A, B, $A + \lambda B$, $A + \mu B$, the symbol of von Staudt corresponding to one arrangement of these points, is effectively $\lambda\mu^{-1}$. We have shewn (p. 25 above) that these points can be placed in perspective with the points B, A, $A + \mu B$, $A + \lambda B$; in accordance with our conventions these last may equally be written $B, A, B + \mu^{-1}A, B + \lambda^{-1}A$. The symbol associated with these four, by the rule suggested, would be $\mu^{-1}\lambda$. The equivalence of the symbols $\lambda\mu^{-1}$ and $\mu^{-1}\lambda$ involves however $\mu\lambda = \lambda\mu$. We have shewn that, unless λ, μ be iterative symbols, that is (p. 86 above) unless the points $A + \lambda B$, $A + \mu B$ are limited to such as are obtainable, from the points A, B and another point of the line taken to be $A + B$, by a finite number of harmonic constructions, the equation $\mu\lambda = \lambda\mu$ in general requires Pappus' theorem.

It is very interesting however to notice that von Staudt recognises the possibility that the symbols which he introduces may not be commutative in multiplication. He gives a proof that the set of four points by which he defines the product of two of his symbols, each defined by a set of four points, is independent of the order in which these two sets are taken (*loc. cit.*, § 20, No. 268, p. 171).

The relation of Pappus' theorem with commutativity of multiplication, under certain hypotheses, is considered by F. Schur, *Analy. Geom., Introd.*, 1898, and by Hilbert, *Grundlagen der Geometrie*, 1899, § 31, p. 71.

CHAPTER II

REAL GEOMETRY

SECTION I. THE PROPOSITIONS OF INCIDENCE. INTRODUCTION OF A PLANE, AND OF A SPACE

Preliminary remarks. The point of view to which we have sought to guide the reader by the preceding discussion is that which we finally adopt as basis of the theory; it appears to possess a simplicity which justifies its being taken first. But it is general, and therefore abstract; as, for instance, in its use of the word *line* in such a sense that every two lines of a plane intersect one another. And it may appear to be artificial; as, for instance, in its adoption of Pappus' theorem. Moreover, it gives no recognition to at least two notions which are, probably, inseparable from any conception of space founded directly on experience. One of these notions is, that space consists of a limited part which is *accessible*, surrounded by an unlimited part which is inaccessible. The other notion, intimately connected with the former, is, that, when a point is given upon a line, there is thereby effected a distinction, between the points of the line on one side of the given point, and those on the other; or, when a line is given in a given plane, there is thereby effected a distinction between the points of the plane, according as they lie on one side, or the other, of the given line; or, when a plane is given in space, a similar separation is thereby made. This notion, applied to the points of a line, leads to the further notion, of the *order* of a set of points on the line. It will be found that the development of this notion of order suggests a view of what points are possible upon a line which, if it were adopted, would make Pappus' theorem inevitable. Though we regard this view as, finally, inadequate, we follow the usual procedure, of allowing the suggestion of a simpler case to become a starting point of the theory of the more general case.

We proceed, then, to examine the fundamental conceptions in a more concrete way than has hitherto been done. Logically, this examination should precede the foregoing formulation; but it will readily be seen that the examination, in proportion as it is complete, is prolix. Moreover, it will appear that the real geometry, which is built up on the basis of this examination, may be regarded as included in the geometry so far dealt with.

Frankly recognising, now, the possibility of inaccessible points, we can assume no general proposition that every two lines in a plane intersect one another, or that any two planes intersect one another ; and, requiring, now, that all constructions made shall utilise only accessible points, we need assumptions as to what points are accessible when others are known to be so. These we obtain, for points of a line, by recognising the separation effected by the assignment of a point of the line; this leads to the notion of the points of the line which are *between* two given points, all of which are regarded as being equally accessible. A similar notion dominates our recognition of the points of a plane which are to be regarded as accessible when three points are given as being so; and similarly for the points of (three dimensioned) space when four points are given.

Betweenness for points of a line. Segments. We assume that when two points, A and B, are given, there exists a point, C, which is *between* A

$$\overline{}$$
$$A' \quad A \quad C \quad B \quad B'$$

and B; and then, also, a point, B', such that B is between A and B'; and also a point, A', such that A is between A' and B.

Thus, between C and B there exists a point, C'; and between A and C there exists a point, C_1.

We assume that the point, C', between C and B, is be-

$$\overline{}$$
$$A' \quad A \quad C_1 \quad C \quad C'\ B \quad B'$$

tween A and B, and the point, C_1, between A and C, is equally between A and B. And we assume, conversely, that every point between A and B, not coinciding with C, is either between A and C or between C and B. Also that B is not between A and C, nor A between C and B. Thence it follows that the point C, not being between C' and B, is between A and C'; and C is, similarly, between C_1 and B.

There is then, for example, a point between C and C', and this point is between C and B, and, therefore, is between A and B. This argument can be continued. There is thus an infinite aggregate of points between A and B; and every point between A and B, not coinciding with C_1, or C, or C', is between A and C_1, or between C_1 and C, or between C and C', or between C' and B.

Again, as B is between A and B', and not between A and C, it is between C and B'. In this statement, C may be any point between A and B. Similarly the point A is between A' and any point which is between A and B.

It appears then, similarly, that there is an infinite aggregate of points B', and an infinite aggregate of points A'.

And, as B is between A and B', the point A is not between B and B'; thus a point A' cannot be the same as a point B'.

Any point, P, between A' and B', and not between A and B, if not coincident with A or with B, is then between A' and A or between B and B'. But there exist points which are not between A' and B', in infinite number, namely points, Q', such that B' is between A' and Q', and also points, P', such that A' is between P' and B'.

We regard the aggregate, of all the points spoken of, as being determined when the points A and B are given. We assume that the same aggregate is determined when any two points of this aggregate are the given points.

When the points A, B are given, and the point C is between them, and B' is a point such that B is between C and B', we say that the points $ACBB'$ are *in order*. If C' is a point between C and B, it is also between A and B, and B is between C' and B', as we have remarked. Thus $AC'BB'$ are in order. These orders are regarded as identical. Again C is between A and C', as we have said, and C' between C and B. Thus $ACC'B$ are in order. This order we again regard as identical with the former. Similarly, A' being, as before, a point such that A is between A' and C, the points $A'ACB$ are in order; and we regard this as identical with the preceding. In this order we then say that B is subsequent to, or follows A, or is superior to A; and that A is anterior to, or precedes B, or is inferior to B. And for greater clearness, we sometimes write this in the form $A < B$. Thus B' follows B, and A' precedes A. Similarly C' follows C, but precedes B.

The order described is determined by the points A, C, B. A similar account may be given of an order determined by the three points B, C, A; this we regard as inverse to the former, speaking of A as, in this order, *following* B, and of A' as following A, of B' as preceding B, and so on.

The aggregate of points C between A and B, together with A and B, we speak of as lying upon the *segment* AB, of which A, B are the end points. The complete aggregate of points determined by A and B, that is, of points C, and of points B' which follow B, and of points A' which precede A, we speak of as lying upon a *line*.

A more minute disentanglement of the ideas involved, than in the foregoing, may be found in the authorities quoted in the Bibliography at the end of this volume.

Points determined by three points. Intersections of lines determined therewith. We now suppose that, beside the points of the line determined by the points A, B, there is given a point D. Each of the three pairs of points, A and B, A and D, B and D, will then be the end points of a segment, and each of these three segments will be part of a line. The point A will not be on the line DB, nor B on DA.

An important step in the argument is then the following: let F

be a point of the segment *DB*, and *G* a point of the segment *AF*: we *assume that there is a point, H, of the segment AB, such that G is*

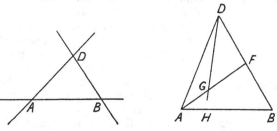

a point of the segment DH. In particular when *F* is *between* *D* and *B*, and *G* is between *A* and *F*, then *H* is between *A* and *B*. For clearness of reference we may often refer to this assumption as Peano's axiom.

From this it follows that, if *F* be between *D* and *B*, and *E* be between *D* and *A*, and *G* be taken between *E* and *F*, then there is a point *K* between *A* and *B* such that *G* is between *D* and *K*. For, applying the axiom to *D*, *E*, *B*, we infer a point *H*, between *E* and *B*, such that *G* is between *D* and *H*; and then, considering *D*, *A*, *B*,

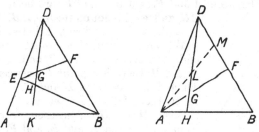

from *H*, between *E* and *B*, we infer *K*, such that *H* is between *D* and *K*. Then as *G* is between *D* and *H*, it is also between *D* and *K*.

And it also follows that, if *F* be between *D* and *B*, and *H* be between *A* and *B*, there is a point *G* which is between *A* and *F* and is also between *D* and *H*. For, if *L* be a point between *D* and *H*, it follows, from the axiom, that there is a point, *M*, between *D* and *B*, such that *L* is between *A* and *M*. If *M* be at *F* we may take for *G* the point *L*. If *M* be not at *F*, it is between *D* and *F*, or between *F* and *B*. Suppose, first, that *M* is between *D* and *F*; then the existence of *L* between *A* and *M*, involves, by the axiom, the existence of a point G_1, between *A* and *F*, such that *L* is between *D* and G_1; and the existence of G_1 between *A* and *F* again involves a point H_1 between *A* and *B*, such that G_1 is between *D* and H_1. Then *L*, between *D* and G_1, is between *D* and H_1. But *L*

is also between D and H. Thus either H and H_1 coincide, or there
is a line determined by H and H_1 which contains both L and D;
in the latter case D would be on the line AB. Therefore H_1 coin-

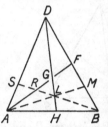

cides with H, and thence G_1, between D and H,
and between A and F, is the point G required.
Next, suppose that M is between F and B;
then, considering A, B, F, the point L, between A and M, involves, as before, a point R,
between A and F, such that L is between B
and R; and, considering D, A, B, the point R,
between A and F, involves S, between A and
D, such that R is between B and S; while, as
L is between B and R, the point R cannot be
between L and B, nor coincide with L, and is between S and L.
Then, from A, L, D, the point R, between S and L, involves a
point G_1, between D and L, with R between A and G_1. However,
R was between A and F. Thus, considering the line determined by
A and R, if we assume that G_1 is between A and F, as it is between
D and L and therefore between D and H, it is such a point G as
was required. And, indeed, considering D, A, M, the point G_1,
between D and L, involves a point, F_1, between D and M, such
that G_1 is between A and F_1; then F and F_1 are both on the line

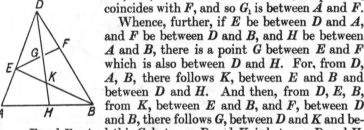

AR, and, as A is not on the line DB, the point F_1
coincides with F, and so G_1 is between A and F.
  Whence, further, if E be between D and A,
and F be between D and B, and H be between
A and B, there is a point G between E and F
which is also between D and H. For, from D,
A, B, there follows K, between E and B and
between D and H. And then, from D, E, B,
from K, between E and B, and F, between D
and B, there follows G, between D and K and be-
tween E and F. And this G, between D and K, is between D and H.
  Also, still with E given between D and A, and F given between

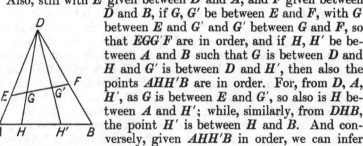

D and B, if G, G' be between E and F, with G
between E and G' and G' between G and F, so
that $EGG'F$ are in order, and if H, H' be between A and B such that G is between D and
H and G' is between D and H', then also the
points $AHH'B$ are in order. For, from D, A,
H', as G is between E and G', so also is H between A and H'; while, similarly, from DHB,
the point H' is between H and B. And conversely, given $AHH'B$ in order, we can infer
that $EGG'F$ are in order.

The statement that when H is between A and B then G is between D and H is the same as that, unless E be at A or F be at B, there is no point of the segment EF which is also a point of the segment AB. Consider now, taking E on the line DA, and F on the line DB, the question whether there are points of the *line EF* which are also points of the *line AB*.

If a point A' be such that A is between A' and B, it follows, by Peano's axiom, considering D, A', B, that, when E is between D and A, there is F, between D and B, such that E is between A and F; and, also, from what we have seen, if B' be such that B is between A and B', there is an F', between E and B', which is between D and B. Thus, when E, F are, respectively, between D and A and between D and B, while there is no

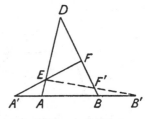

point of the segment EF which is a point of the segment AB, there *may* be a point, A', or B', of the *line AB* which is a point of the *line EF*.

We may examine the various combinations of the possible positions of E on the line DA, with those of F on DB, putting together those in which such a common point of the two lines AB, EF necessarily follows from Peano's axiom, and then those for which this does not follow. There are evidently nine possible combinations.

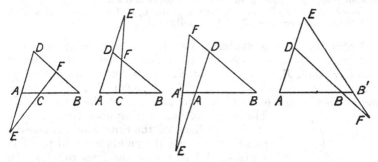

When F is between D and B and E not between D and A, whether A be between D and E, or D be between A and E, there is necessarily a point, C, of the line EF which is a point between A and B. And similarly when E is between D and A. This accounts for four combinations. When F is not between D and B, it may be such that D is between F and B; then if E be such that A is between E and D, there is a point A' between E and F such that A is between A' and B; while, if F be such that B is between D and F, and E be such that D is between A and E, there is a point B'

between E and F such that B is between A and B'. This accounts for two other combinations. For the three remaining combinations no common point of the two lines is certain, the last of these being the case considered above. In these three cases, E, F are either

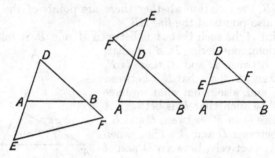

both interior points of the segments DA, DB, or neither is an interior point. In none of these three cases can there be a common point of the segments AB and EF, as we easily prove from the axiom.

In any case, the *total number of points of the line EF which lie within one of the three segments AB, DA, DB, is zero or two.*

Example. If E, F be points such that D is between A and E, and D is between B and F, and G be any point between E and F, prove that there is a point C between A and B such that D is between C and G.

Considering E, F, B, there follows a point, L, of DE, between B and G, and, then, considering A, B, G, there follows C.

Points of a plane system. Let A, B, C be three given points, not belonging to one line. A line is determined by the point A and any point, P, of the *segment BC*; so, a line is determined by the point B and any point, Q, of the segment CA, and a line is determined by the point C and any point, R, of the segment AB.

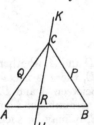

The system of points lying upon the aggregate of all such lines, of the three kinds, is called a plane system, and the points are said to lie in a plane. All points of the lines BC, CA, AB belong to this system. The system contains points lying upon a line of all the three kinds, AP, BQ and CR; but, also, it contains points lying upon a line of one kind only.

Consider a line of the kind CR, R being between A and B, and, thereon, points, H and K, such that R is between H and C, and C is between R and K. A point between C and R is equally upon a line AP, and upon a line BQ, by what we have shewn; and, therefore, neither H nor K is upon

a line of the kind AP or the kind BQ. If B' be a point of the line
AB such that B is between A and
B', a point, M, between C and B'
is on a line AP; a point, H, such
that B' is between C and H, is
upon a line BQ, as also is a point,
K, such that C is between K and
B'. A similar remark can be
made, starting from a point A'
such that A is between A' and B.
The plane system thus contains
the points of all lines determined

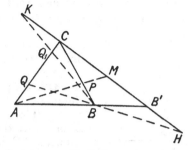

by C and a point of the *line AB*; but to say that the system con-
sisted only of such points would be to assume that every line, in
the plane, drawn through C, had a point of intersection with the
line AB.

Let E be a point between C and A, and F a point between C
and B; and let H, K be points of the line EF,
the point F being between E and H and the
point E between K and F. It is clear from
what has been said that a point of this line
between E and F lies on a line CR, where R
is between A and B. Consider C, A, H; as E
is between C and A, and F is between E and
H, there is a point, T, between A and H, such

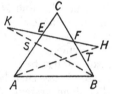

that F is between C and T; thus H lies on a line through A containing a point of the line CB. We remarked that the plane system
contains all points of a line determined by C and a point of the
line AB. By similar reasoning it contains all points of a line
determined by A and a point of the line CA. Thus H is a point
of the system. Again consider C, K, B; as F is between C and B,
and E is between K and F, there is a point, S, between K and B,
such that E is between C and S; thus K lies on a line through B
containing a point of the line CA. It is thus shewn that every
point of the line EF belongs to the plane system. Therèfore a line,
containing points of any two of the segments BC, CA, AB, belongs
to the plane.

Now let A', B', C' be any three points of the plane system not
lying on one line. We shew that these may be used, instead of
A, B, C, to determine the plane system. For this it is sufficient to
prove that if H be a point such that at least one of the three possibilities, that the line HA' contains a point of the segment $B'C'$,
or the line HB' contains a point of the segment $C'A'$, or the line
HC' contains a point of the segment $A'B'$, is a fact, then at least
one of the corresponding possibilities, for H in regard to A, B, C,

is also a fact, and to prove that the converse is also true. This proof will follow if we shew that, when A' is any point of the plane system (A, B, C), the plane systems, (A', B, C) and (A, B, C), are the same.

Now, first, if A' be between A and C, a point, H, for which the line HA' contains a point between B and C, is on a line containing points of two of the segments AC, CB, BA; so that such point H belongs to the plane system (A, B, C), by what was proved above. The same follows if the line HA' contains B, or C. If the line HB contains a point of the segment $A'C$, it contains a point of the segment AC; and if the line HC contains a point of the segment BA', it contains a point of the segment AC; in either case it belongs to the plane system (A, B, C). Conversely, if the line HA contains a point of the segment BC, other than B, it contains a point of the segment BA', and so belongs to the plane system (A', B, C), by what was proved above; and the same follows when H is on the line AB. Again, if the line HB contains a point of the segment CA, it contains a point of the line CA', and belongs to the system (A', B, C); while, if the line HC contains a point of the segment AB, it contains a point of the segment $A'B$, and then, also, belongs to (A', B, C). A point between any two of the points A, B, C may then be used instead of either of these to determine the plane system. The plane systems so obtained are the same. But from this it follows that any one of the points, A, may be replaced by any point on either of the *lines*, AB, AC, containing this point.

Now let A' be any point of the plane system (A, B, C), not lying on the line BC: then, either

(*a*) the line BA' contains a point of the segment CA, say the

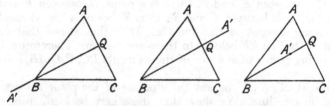

point Q; then the plane system (A', B, C) is the same as (Q, B, C) —each being the same as (Q, A', C)—and (Q, B, C) is the same as (A, B, C), or

(*b*) the line CA' contains a point of the segment BA, in which case a similar argument shews that the systems, (A', B, C), (A, B, C), are the same, or, lastly,

(c) the line AA' contains a point of the segment BC, say the

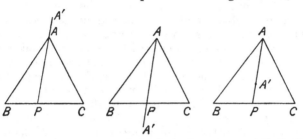

point P; then the system (A', B, C) is the same as (A', B, P), and this as (A, B, P), and this as (A, B, C).

It is thus established, whatever be the point A' of the plane system (A, B, C), not lying on the line BC, that the system (A', B, C) is the same as (A, B, C). Hence it is easy to deduce that the system (A, B, C) may be defined from any three points (A', B', C') of itself which are not in line.

An incidental consequence is that all the points of the line determined by any two points of a plane system lie also in the system.

Separation of the points of a plane effected by a line therein. Let us say that two points of a plane system belong to the same class when the segment of which they are the end points contains no point of a certain line lying in the plane, but that two points belong to different classes in the contrary case. Then the results we have developed shew that every point of the plane belongs to one of two classes, save that the points of the chosen line are exceptional. For let A, B be any two chosen points of different class, the segment AB containing a point of the line; let C be any other point of the system not lying on the chosen line: we have shewn that, of points of the three segments AB, BC, AC, there are two or none lying on a line of the plane: the chosen line therefore contains a point of the segment AC, or a point of the segment BC. The point C thus belongs either to the same class as A or to the same class as B. Moreover, in this description, A may be replaced by any point of its own class, and B by any point of its class. For if A' and C be of the same class as A, the line not containing a point of either segment AA' or AC, then the line does not contain a point of the segment $A'C$, and C belongs to the same class as A'. Similarly if B' be of the same class as B, but C of a different class from B, then C, B' are of different class.

As soon as we are in a position to shew that we can do so in a self-consistent manner, we shall discard the distinction between the accessible and the inaccessible points of a plane, upon which our

theory at present rests; and then the separation of the points of a
plane, effected by a line, will disappear, unless reintroduced by a
convention as to which points are to be regarded as accessible. But
for the present this separation is most important. If it were introduced at the start, as an axiom, and three lines were taken in the
plane, not having a common point; and the two classes of points
of the plane with reference to one line were denoted, say, by α and
α', the two classes with reference to another of these lines by β and
β', and the two classes with reference to the remaining line by γ
and γ', then the points of the plane, not lying on any of the lines,
would be each of one of seven categories $(\alpha, \beta, \gamma), (\alpha', \beta, \gamma), (\alpha, \beta', \gamma)$,
$(\alpha, \beta, \gamma'), (\alpha, \beta', \gamma'), (\alpha', \beta, \gamma'), (\alpha', \beta', \gamma)$. Peano's axiom would then
arise in a very natural way.

Example. Given two lines, AB, AC, with a common point, A, it may be
proposed to define a plane system as the aggregate of points through which a
line can be drawn having a point, other than A, common with each of the
two given lines. Examine whether this system is coextensive with that,
adopted above, defined by the points A, B, C.

**Space systems, defined by four points not lying in a
plane.** As a preliminary remark, before proceeding to a formal
definition, suppose that we have a system of points, not lying in a
plane, of such character and multiplicity that it contains an infinite
number of plane systems, while those of the system not lying on a
specified plane are, by means of this plane, divided into two classes.
Then, if A, B, C, D be four points of the system not lying in a
plane, the plane determined by A, B, C distinguishes two classes,
say δ and δ'; similarly the plane BCD distinguishes two classes,
α and α'; the plane CAD two classes, β and β'; and the plane ABD
two classes, γ and γ'. The points of the system are then, by these
four planes, distinguished into $2^4 - 1$, or fifteen, categories, such as
$(\alpha, \beta, \gamma, \delta)$, or $(\alpha', \beta, \gamma, \delta)$, etc., exception made of points on one
of the four planes. Suppose further that two points of the system
are of the same class, with respect to one of the planes, or of different class, according as the segment of which these are the end
points does not, or does, contain a point of this plane, it being
understood that, with reference to the plane ABC, the point D is
of class δ, while A is of class α with reference to the plane BCD,
and B of class β with reference to the plane CAD, and similarly
for C. Then, if a point of the system, P, be of the category
$(\alpha, \beta, \gamma, \delta')$, the segment PD, whose end points are respectively of
classes δ' and δ, contains a point of the plane ABC; this will,
however, be of category (α, β, γ) with reference to the other three
planes, this being so both for P and D. A point P of the category
$(\alpha, \beta, \gamma, \delta')$ may therefore be said to be such that the line PD contains a point of the *face* BCD; this statement is then equally true

for a point of the category $(\alpha', \beta', \gamma', \delta)$. But if a point, P, be of either of the categories $(\alpha, \beta', \gamma', \delta)$ or $(\alpha', \beta, \gamma, \delta')$, then a line can be drawn through P to contain both a point of the segment BC and a point of the segment AD. And similarly in other cases. We therefore say:

Let A, B, C, D be four points not lying in a plane; let the points, Q, of the plane ABC which are such that, simultaneously, the line AQ contains a point of the segment BC, the line BQ contains a point of the segment CA and the line CQ contains a point of the segment AB, be said to belong to the *face ABC*; with a similar definition of the faces BCD, CAD, ABD. Then a (threefold) space system, or, briefly, a space, is the aggregate of all possible points, P, which are such that, either the line PD contains a point of the *face ABC*, or the line PA of the face BCD, or PB of CAD, or PC of ABD, or all of these, together with points, P, such that a line is possible containing P which contains a point of each of the segments BC and AD, or a line, containing P, which contains a point of each of the segments CA and BD, or a line, containing P, which contains a point of each of the segments AB and CD. To this space belong all the points of the four planes ABC, BCD, CAD, ABD.

In a manner analogous to that employed for plane systems it can then be deduced

(a) That, if A', B', C', D' be any four points of the space system which do not lie in a plane, the system can equally be determined by these. In particular the points of the line determined by any two points of the system, or the points of the plane determined by any three points of the system not lying in line, all belong to the space system.

(b) That, when a plane of the space is given, the points of the space system are divided into two classes, two points, E, F, being of the same or of different classes, according as there is no point of the segment EF, or there is a point, which belongs to the plane.

(c) That any two planes of the space which have a point in common have all the points of a line in common.

(d) That, if P, Q, R be three points of the space system which are not in line, there are none, or two, points of any plane, belonging to the space, which are points of the segments QR, RP, PQ.

Of these, (a) is proved by first proving that D can be replaced by any point on any of the lines DA, DB, DC, and continuing this process; then (b), from the definition of the space system, is seen to be true for any one of the planes ABC, BCD, CAD, ABD; after (a) it is then true for any plane. For (c), if α, β denote the two planes, and O their common point, take two points P, P' in the plane α such that O is between them. As O belongs to the plane β, the points P, P' are of different class with respect to β. Now let

Q be another point of the plane α, not lying on β; of the two points P, P', let P be that one which is of the same class as Q with respect to the plane β, so that P', Q are of different class. The segment $P'Q$ thus contains a point of the plane β, say O'; this point, lying on the segment $P'Q$, is also on the plane α. Thus both the planes contain the two points O and O', and the line determined by these is, therefore, common to the two planes. The statement (d) then follows by remarking that, in virtue of (c), if a plane contain a point of one of the segments QR, RP, PQ, it meets the plane PQR in a line. For we have shewn that in a plane (PQR) a line contains none, or two, points of the segments QR, RP, PQ.

Extension to fourfold and higher space. If now it be allowed that five points A, B, C, D, E can exist, of which A, B, C, D are not in a plane, such that E does not belong to the threefold space system determined by A, B, C, D, we can regard these five points as determining a fourfold space, specifying, in a manner analogous to that followed above for twofold and for threefold space, what points are to be regarded as belonging thereto. Such a space will contain lines, and planes, and also threefold space systems. It will not be true, in such a fourfold system, to take one illustrative example, that two planes which have one point in common, necessarily have in common the points of a line; for it will now be by a threefold space, and not by a plane, that the points of the fourfold system are separated into two classes. But a similar proof to that followed above will shew that two threefold spaces of the fourfold system which have a point in common, have also a plane in common; let α, β denote the threefold spaces, O the assumed common point, and P, P' two points of α having O between them; as before a point Q is possible in α such that the segment QP' contains a point, O', of β which also lies in α, so that the line OO' is common to α and β. But now, beside Q, a point R is possible, in α, which does not lie in the plane QPO; and this gives a further point, O'', common to α and β, which does not belong to the line OO'. The points O, O', O'' determine a plane common to α and β.

We regard such a fourfold space, or spaces of higher multiplicity, as entirely possible; and experience has shewn the great usefulness and interest of taking them into consideration. It seems, however, unnecessary, at this stage, to enter into further detail.

SECTION II. THE GENERALISATION OF THE REAL GEOMETRY
BY THE INTRODUCTION OF POSTULATED POINTS

Purpose of the Section. In comparison with the geometry of Chapter I, the Real Geometry we have dealt with in the preceding Section is inconvenient to apply, owing to such circumstances as that two lines in a plane may not have a common point, and two planes in a threefold space may not have a common line. The common point of two lines, in Chapter I, is, however, only of use because, as a point, it can serve as one of two points determining a line; in other words, the proposition of Chapter I is that, if two lines be given in a plane, there is thereby determined through an arbitrary point which does not lie on either of them, a definite line having no common point with either of them, save perhaps where they have a common point. Now, in this form, the proposition is equally true for the Real Geometry. When this is made clear it will be convenient to speak of the two given lines as having in common a so-called *postulated point*, leaving open the question whether this point is accessible or not. Similar remarks apply to the other cases in which the Real Geometry appears incomplete in comparison with that dealt with in Chapter I. The result of the whole examination now to be made is that, we may regard the space of the Real Geometry as part of a space in which the Propositions of Incidence formulated in Chapter I are completely valid, provided that postulated points, lines and planes are allowed an existence, whether they be accessible or not.

Extended form of Desargues' Theorem, for lines and planes having a common point. Let O be a point, and a, b, c, a', b', c' be six lines passing through this point, no three of which lie in a plane. The pairs, a, a', and b, b', and c, c', of these lines, will determine three planes containing the point O; any two of these planes will, therefore, have a line in common, passing through O. Suppose the three lines so arising coincide with one another, so that the three planes all contain one line, l, passing through O. There are six other planes determined by the given lines, of which two are the planes containing respectively the pairs b and c, and b' and c'; these two planes, which we may respectively call $[b, c]$ and $[b', c']$, meet in a line through O. There are then two other lines through O, given, respectively, by the planes $[c, a]$, $[c', a']$ and $[a, b]$, $[a', b']$. With the hypothesis made, that the three planes $[a, a']$, $[b, b']$, $[c, c']$ have a line, l, in common, we prove that the three lines which are the intersections of the respective pairs of planes

$$[b, c], [b', c']; \quad [c, a], [c', a']; \quad [a, b], [a', b'],$$

lie in one plane. This plane will pass through O.

To prove this, remark, first, that if three lines a, a', l in a plane have a point O in common, and points P', Q' be taken on a' so that O is between P' and Q', the line l must have a point in common

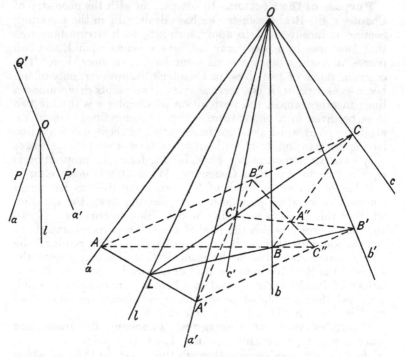

either with PP' or PQ'. For, as we saw in Section I, the line l has none or two points, in all, lying on the segments PP', $P'Q'$, $Q'P$; and it has the point O on the segment $P'Q'$. It is thus possible to take P' and P (or Q' and P), respectively on a' and a, so as to lie on opposite sides of the line l.

It follows that, in our figure, points A, A' can be taken, respectively on the lines a and a', such that there is a point, L, of the line l, lying between them. By a similar argument, a point, C, can be taken, on the line c, so that there is a point, C', of the line c', lying between L and C. Then it follows, from Peano's axiom, that, in the plane $AA'C$, there is a point, B'', lying between A and C, such that C' is between A' and B''. Again, a point, B', can be taken, on the line b', so that there is a point, B, of the line b, lying between L and B'; then it follows, from Peano's axiom, that, in the plane $AB'A'$, there is a point, C'', between A' and B'', such that B is between A and C''. It follows further, by the axiom, that, in the plane CLB', there is a point, A'', which is between C' and B' and also between C and B.

It is at once seen that the points A'', B'', C'' are upon the plane ABC and upon the plane $A'B'C'$. Thus A'', B'', C'' are in line. But the line OA'' is the line of intersection of the planes $[b, c]$ and $[b', c']$; similarly OB'' is the line of intersection of the planes $[c, a]$ and $[c', a']$, and OC'' is the line of intersection of the planes $[a, b]$ and $[a', b']$. These three lines of intersection are therefore in one plane; as we desired to prove.

Consider now the converse proposition. As before consider six lines a, b, c, a', b', c', passing through the point O, of which no three are in one plane; suppose that the lines of intersection of the respective pairs of planes, $[b, c]$ and $[b', c']$, $[c, a]$ and $[c', a']$, $[a, b]$ and $[a', b']$, are in one plane. It can be shewn that the planes $[a, a']$, $[b, b']$, $[c, c']$ have a line in common.

For, let the planes $[a, a']$, $[b, b']$, which meet in O, meet in the line l. Let the line of intersection of the plane $[l, c]$ with the plane $[a', c']$ be the line c_1, passing through O; so that the planes $[a, a']$, $[b, b']$, $[c, c_1]$ have the line l in common. Then, by the preceding, it follows that the lines of intersection of the three pairs of planes $[b, c]$ and $[b', c_1]$, $[c, a]$ and $[c_1, a']$, $[a, b]$ and $[a', b']$, lie in one plane. By construction, however, the plane $[c_1, a']$ is the same as the plane $[c', a']$; thus, of the three lines which are in one plane, two are the intersections, respectively, of $[c, a]$ and $[c', a']$ and of $[a, b]$ and $[a', b']$; and it was to be supposed that the plane of these contains the line of intersection of the planes $[b, c]$ and $[b', c']$. The line of intersection of the plane of these, with the plane $[b, c]$, thus lies in both the planes $[b', c_1]$ and $[b', c']$; and, therefore, these last planes, both containing the line b', coincide with one another. Wherefore the lines c_1, c' are the same, or the plane containing them contains the line b'. By construction, however, the plane of c_1 and c' contains the line a'; and, it was assumed that the lines a', b', c' are not in one plane. Thus it follows that the lines c_1 and c' coincide; and, therefore, the planes $[a, a']$, $[b, b']$, $[c, c']$ meet in a line, as was to be proved.

For the proposition here given, and its application, we may refer to Reyes-y-Prosper, "Sur les propriétés graphiques des figures centriques," *Math. Annal.* XXXII, 1888, 157; Pasch, *Math. Annal.* XXXII, 1888, 159; F. Schur, *Grundlagen der Geometrie*, 1909, § 2.

Fundamental theorem in regard to the correspondence of the lines and planes of two central figures, or stars. A figure consisting of lines and planes passing through a point, such as that considered above, may be called a *central figure*, or a *star*; the point itself being called the centre. Suppose now we have two centres, O_1 and O_2. Let α be a plane, not passing through O_1 or O_2.

To every line through the point O_1 which has a point in common with the plane α, we can evidently make correspond a line through the point O_2, namely that which meets the plane α in the same point as does the former; and either of these lines determines the other. To establish such a correspondence it is not necessary, however, to assume an actual, or accessible, intersection of the former line with the plane α. We can shew that, *to every line through the centre O_1, there corresponds a definite line through the centre O_2, and conversely, without utilising an actual, or accessible, point common to the plane α and the lines. And this in such a way that to three lines through O_1 which lie in a plane, there shall correspond three lines through O_2 which also lie in a plane—these planes coinciding when the former plane, through O_1, passes through O_2.*

First consider the statement in regard to single lines through O_1 and O_2. Let l_1 be any line through O_1; we make no use of any intersection it may have with the plane α. Upon the plane α take three arbitrary points; through the line l_1 pass three planes, each containing one of these points; any one of these planes, having a point common with the plane α, will have a line of intersection with it. Denote these lines by a, b, c, respectively. If the line l_1 has an actual point of intersection with the plane α, each of the lines a, b, c will pass through this point; conversely a point, common to two of the lines a, b, c, lies upon two planes passing through the line l_1, and therefore lies upon l_1, as well as upon the plane α. Thus, to refrain from using any possible intersection of the line l_1 with the plane α, we must avoid the assumption of an accessible intersection

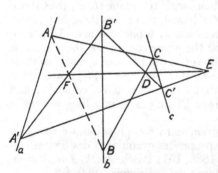

of any two of the lines a, b, c. On the line a, take two arbitrary points, A and A'. By what has been seen above in regard to the separation of the points of a plane effected by a line of that plane, it is possible to take a point, E, of the plane α, such that each of the segments EA and EA' contains a point of the line c. Say these points are, respectively, C and C'. On the line b take an arbitrary point, B; it is then possible to take a further point, B', upon the line b, such that there is a point, D, of the segment BC lying between B' and C'. Then, considering the points A', C', B', we see that there is a point F between A' and B', such that D is between F and E.

We can now prove, by the theorem of the foregoing article, that the points A, F, B are in line. For, let the lines joining the point O_1, respectively to the points A, B, C, A', B', C', be called $a_1, b_1, c_1,$ a_1', b_1', c_1'; so that the planes $[a_1, a_1'], [b_1, b_1'], [c_1, c_1']$ are, respectively, the planes through O_1 containing the lines a, b, c. By construction these planes have a line in common, the line l_1. Therefore, by the theorem referred to, the lines of intersection of the three pairs of planes

$$[b_1, c_1] \text{ and } [b_1', c_1'], \ [c_1, a_1] \text{ and } [c_1', a_1'], \ [a_1, b_1] \text{ and } [a_1', b_1']$$

lie all in one plane; of these lines, however, the two former are respectively the lines O_1D and O_1E; and the last line meets the plane α in a point lying upon both the lines AB and $A'B'$. Thus the point F must lie on the line AB.

By the converse part of the theorem of the foregoing article, we can thence infer that the planes through O_2 which contain, respectively, the lines a, b, c, also meet in a line, which we shall call l_2. For, if the lines joining the point O_2 to the points A, B, C, A', B', C' be, respectively, called $a_2, b_2, c_2, a_2', b_2', c_2'$, it follows from the fact that the lines which are the intersections of the three pairs of planes

$$[b_2, c_2] \text{ and } [b_2', c_2'], \ [c_2, a_2] \text{ and } [c_2', a_2'], \ [a_2, b_2] \text{ and } [a_2', b_2'],$$

lie in a plane, namely the plane through O_2 which contains the line DEF, that the three planes $[a_2, a_2'], [b_2, b_2'], [c_2, c_2']$ meet in a line. These are the planes joining O_2 respectively to the lines a, b, c.

It is thus proved that, given the points O_1 and O_2, and a plane α not containing O_1 or O_2, if any line l_1 be drawn through O_1, and, through this, be taken three planes intersecting the plane α respectively in the actual lines a, b, c, then the planes through O_2 containing these lines a, b, c have in common another line, l_2, which will pass through O_2. There is thus determined a line through O_2 corresponding to any given line passing through O_1; and, if the line through O_2 be given, that through O_1 is determined by the same construction. The lines a, b, c were determined by planes containing l_1, drawn, respectively, through three arbitrarily taken points of the plane α; it is easy to see that the line l_2 is the same whatever these points may be. For, if two of these points remain the same, say those which lead to the lines b and c, the third may vary; the line l_2 will still be determined by the intersection of the planes joining O_2 to the lines b and c. By varying, thus, one point at a time, all can be varied. The construction given has not used any intersection of the plane α with the line l_1; when such a point is given the line l_2 is that joining O_2 to this point.

Next consider three lines, passing through O_1, which lie on a plane; we desire to prove that the lines through O_2, corresponding to these respectively, by the above rule, also lie in a plane. Let the given lines through O_1 be p, q, r. Take a point P of the plane α; the plane, Pr, containing the line r and the point P of the plane α, will meet α in a line; let Q be a point of this line other than P. Let the lines O_1P and O_1Q be called, respectively, a and b. The planes Pq and Qp, that is the planes joining the points P and Q respectively to the lines q and p, have the point O_1 in common; they therefore have a line in common, which we call c. Similarly, take another point P' of the plane α, and then, in the line in which the plane $P'q$ meets the plane α, take another point R'; denote the lines O_1P' and O_1R', respectively, by a' and c', and denote the line of intersection of the planes $P'r$ and $R'p$ by b'. Then

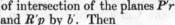
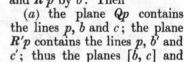

(a) the plane Qp contains the lines p, b and c; the plane $R'p$ contains the lines p, b' and c'; thus the planes $[b, c]$ and $[b', c']$ meet in the line p;

(b) the plane Pq contains the lines q, c and a; the plane containing q and $P'R'$ contains the lines q, c' and a'; thus the planes $[c, a]$ and $[c', a']$ meet in the line q;

(c) the plane $P'r$ contains the lines r, a' and b'; the plane containing r and PQ contains the lines r, a and b; thus the planes $[a, b]$ and $[a', b']$ meet in the line r.

It was said, however, that the lines p, q, r are in one plane. Thus, by the theorem of the last section, we are able to infer that the planes $[a, a'], [b, b'], [c, c']$ have a line in common. Call this line l_1. Further these planes have each a line common with the plane α, having, respectively, thereon, the point P (upon a), the point Q (on b) and the point R' (on c'). Whence, by what has immediately preceded, the planes joining the point O_2, respectively to these lines of the plane α, have also a line in common, say l_2. And, also, to the six lines a, b, c, a', b', c' through O_1, will correspond, respectively, lines through O_2, say $a_2, b_2, c_2, a_2', b_2', c_2'$, there being through O_2, corresponding to the planes $[a, a'], [b, b'], [c, c']$, the planes $[a_2, a_2']$, $[b_2, b_2'], [c_2, c_2']$. As these meet in a line, l_2, the lines of intersection of the three pairs of planes

$$[b_2, c_2] \text{ and } [b_2', c_2'], \quad [c_2, a_2] \text{ and } [c_2', a_2'], \quad [a_2, b_2] \text{ and } [a_2', b_2'],$$

are three lines, through O_2, lying in a plane. These are the lines, say p_2, q_2, r_2, through O_2, which correspond, respectively, to p, q, r. We have thus proved what we desired to prove. We have not utilised any intersections of p, q, r with the plane α; if such points exist, they will lie, respectively, on p_2, q_2, r_2.

Lastly, we prove that, in this manner, to a plane, ϖ, through O_1 which contains O_2, corresponds, as plane through O_2, this plane itself. The result of course is obvious if the plane ϖ meets the plane α, any line through O_1 in the plane ϖ corresponding then to a line through O_2 in this plane meeting the plane α in the same point as the former; but, as before, we can avoid using the possible intersection of the plane ϖ with the plane α.

When there is a point of the plane α between O_1 and O_2, any plane through O_1O_2 will intersect the plane α in a line, and any line through O_1 drawn in such plane to contain a point of the plane α on this line, will correspond to a line through O_2, lying in such plane, meeting the plane α in the same point as does the line to which it corresponds. In this case the argument can be simplified; it will then be sufficient to suppose that there is no point of the plane α lying between O_1 and O_2. We can then take a point O_3 so situated that the plane α contains a point of both the segments O_1O_3 and O_2O_3; the simplifying considerations are then effective, for example, for O_2 and O_3.

To any line, l_1, drawn through O_1, we can make correspond, as above, with respect to the plane α, a line, l_2, through O_2, and, also, a line, l_3, through O_3. Take a point, H, between O_1 and O_2, so that the plane Hl_3, through H and l_3, containing a point of the plane α, meets α in a line, say h. We can then prove that the line l_2 lies in both the planes O_2l_3 and O_2h; it lies in the plane O_2l_3, because this plane meets α in a line, of which any point, joined to O_2 and O_3, gives corresponding lines, while, as we have proved, lines through O_2 lying in a plane correspond to lines through

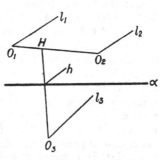

O_3 also lying in a plane; it lies in the plane O_2h, because it must lie in the plane through O_2 corresponding to the plane Hl_3 through O_3, which contains l_3, and this latter plane meets the plane α in the line h.

Now consider correspondence, not with respect to the plane α, but with respect to the plane Hl_3; it will be seen that the line through O_2 corresponding to l_1 in this way, is also l_2. For just as l_2 was shewn to lie in the plane O_2h, so it follows that l_1 lies in the plane O_1h. The line through O_2 corresponding to l_1, with respect to the plane Hl_3, must therefore lie in the plane O_2h, the line h being in the plane Hl_3; for the same reason it must be in the plane O_2l_3. These two facts shew that this line is l_2.

But then, as l_1, l_2 correspond to one another with respect to the plane Hl_3, which contains a point H between O_1 and O_2, they lie in a plane through O_1 and O_2. And, then, as l_1 is an arbitrary line through O_1, there follows what was desired, that, to any plane through O_1 containing O_1O_2, there corresponds, with respect to the plane α, as plane through O_2, the same plane.

Postulated points. Consider any two lines in a plane, of which the intersection, if it exists, is not given. With the help of the foregoing considerations, we can shew that it is possible to make constructions which are of the same effect as if the point of intersection were known to exist and were accessible and given; and this in a unique manner. This will enable us to speak of the lines as having what we may call a *postulated point* of intersection.

Let the lines be a, b, lying in the plane α. Let O be any point, not lying in this plane. The two planes Oa, Ob, with the common point O, have a line in common, passing through O, which we call l. Let c be a third line of the plane α, such that the plane Oc also contains l. We desire to find justification for saying that the lines a, b have a postulated point of intersection, and that the line c passes through this. For this to be legitimate it is sufficient, first, that the statement shall not be dependent upon the position of the point O; the preceding article shews that it is not; and, second, that if the lines a, b have an actual point of intersection, the line c shall contain this, which is obvious.

Then, if C be an arbitrary point of the plane α, the line from C to the postulated point of intersection of the lines a and b, would, by the definition, be the line of intersection, with the plane α, of the plane Cl. As the preceding article shews, the construction for this line may be made in the plane α alone; we have only to take two arbitrary points, A, A', on the line a, an arbitrary point, B, on b, then a point, B', on b such that a point, F, between A and B, is between A' and B'; then, further a point, E, so that the given point C is between A and E, so that, from the points A, B, E, there is a point, D, between B and C which is also between F and E, and there is also, from the points A', B', E, a point, C', between A' and E, such that D is between B' and C'. The line CC' is then the line required. The line which we regard as determined by the postulated point, and a point O not lying in the plane of a, b, is the line, l, found as above, from the planes Oa and Ob.

The postulated point has then, like an actual point, the capability of determining a line, when taken with another point. It is likewise sufficient, when taken with two other points, O and O', to determine a plane; for we have seen, regarding the postulated point as arising from two given lines a, b, lying in a plane α, that these, with the point O, outside this plane, give rise to a line, l, passing through O; with the point O', they also give rise to a line, l', passing through O'; and we have seen that these corresponding lines, through O and O', lie in a plane. This plane, in virtue of its construction, may be said to contain the postulated point.

If we have two pairs of lines in the plane α, namely a, b and a', b', but no actual intersection of a, b, or of a', b', we have, thereby, two postulated points. With a point O, not lying in the plane α, these two points determine a plane; for they determine, by the preceding, two lines through the point O, say the lines l and l'. The plane of these is the plane in question.

When we have three postulated points of the plane α, it may happen that the lines l, l', l'' similarly arising, one from each of these points, through a point O, not in the plane α, lie in one plane. By the preceding article it follows, that, if this is so for one position of the point O, it is so for another position. We shall then be justified in speaking of the three postulated points as being in line. This leads to a new conception, that of the *postulated line*, which is defined in the first instance by two postulated points; if the postulated points be actual, the postulated line which joins them is also actual, and subject to the construction given.

The postulated point may equally be defined by means of a plane, α, and a line, l, not lying in this plane; the line joining any given point, P, of the plane α, to the postulated point, being the intersection of the plane Pl with the plane α, this definition accords with the former. And, similarly, the postulated line joining two postulated points of the plane α, is, by the preceding, in fact determined by two planes, the plane α and another. Conversely, if two planes, say α and ϖ, be given, but no intersection of these, we can find postulated points by the join of which a postulated line is determined which we may regard as determined by the planes α and ϖ; for take on the plane α the point A, and on the plane ϖ the point O; then, an arbitrary plane through the line AO meets the plane ϖ in a line, say l, and meets the plane α in a line, say a; if B be another point of the plane α, the plane through l and B meets the plane α in another line b; these lines a, b, in the plane α, define a postulated point of this plane which is equally, in accordance with what precedes, to be regarded as being on the plane ϖ. It is also clear incidentally that, a postulated line determines, with an actual point, an actual plane.

Summing up the results so far obtained, we may then say:

1. Two actual lines in a plane have a point of intersection, actual or postulated.

2. A line containing an actual point is an actual line; for two planes having an actual point common have an actual line common. But a point lying on an actual line may quite well be a postulated point.

3. An actual line and a postulated line, in the same plane, have a point in common, which is a postulated point. If the postulated line be that joining two postulated points defined respectively by the pairs of actual lines a, b and a', b', lying all in a plane, the actual line being the line, c, of the same plane, and we take a point, O, external to this plane, the planes Oa, Ob intersect in a line l, and the planes Oa', Ob' intersect in a line l'; and the point common to the line, c, and the postulated line is that defined by c and the plane (l, l').

4. Similarly it may be seen that two postulated lines in a plane have a point common, a postulated point. More generally, from an actual point an actual line can be drawn to meet in a point, actual or postulated, each of two skew lines; for an actual plane can be drawn through the point to contain a line, actual or postulated, and two such planes through the point have an actual line in common.

5. We have seen that an actual line and an actual plane not containing this line, define a point, actual or postulated. More generally a postulated line and an actual plane define a point, unless the line lie in the plane; for, from an arbitrary point of the plane, there passes a line to each of two points taken on the line, and these two lines determine a plane meeting the given plane in a line; this line with the postulated line determine the point in question.

6. Thus, also, any three actual planes meet in a point, actual or postulated.

7. The duality between actual lines and points, in a plane, continues to hold for postulated points and lines, being a consequence of the reciprocity existing for a star; not only is it true that two lines through the centre of the star determine a plane, but, also, two planes through the centre of the star meet in a line. In three dimensions, a line and a plane determine a point, which may be a postulated point; the reciprocal theorem would be that a line and a point determine a plane. If the point, or the line, be only postulated, we may still regard them as determining a plane; but in case both the point and line are postulated elements this is a *postulated plane*. It must then be proved that the line determined by two points of a postulated plane lies entirely in that plane; and, when the notion of a postulated plane is acquired, it is to be shewn that two postulated planes determine a line, which will be a postulated line because every plane through an actual line is an actual plane; and, further, that three postulated planes determine a (postulated) point, and that a line and a plane in all cases determine a point. It seems unnecessary to develop these results in detail; the above account of the course of the argument will make its character clear.

Remark on the theory of postulated elements. A particular case of the foregoing theory is that view of Euclid's Geometry wherein two parallel lines are said to have a point in common, lying at infinity. This view is at least as old as Kepler (*Ad Vitellionem paralipomena, de Coni sectionibus*, Francofurti, 1604), who deduced it independently of notions of perspective, and Desargues (*Brouillon project.*, Paris, 1639; *Oeuvres de Desargues, par M. Poudra*, Paris, 1864; t. I, pp. 104, 243. See Ch. Taylor, *Ancient and Modern Geometry of Conics*, Cambridge, 1881, pp. lvi–lxxii). In this view every line has a single postulated point, and two lines are parallel when they intersect therein; if two other lines both pass through the postulated point of a line, this is the postulated point of all. In the more general theory outlined above, two lines, *b*, *c*, may determine different postulated points of a line *a* but intersect one another in an actual point. We arrive at such a theory, in a plane, by confining our attention to a limited portion of the plane, regarding as actual, points in this limited portion, and, as postulated, points without this. The theory involves the definite discovery that we can make constructions affecting the points exterior to the limited portion though only using *as data* the points interior to it. The limited portion of the plane which is considered must be such that the segment, whose end points are any two points of this, contains only points belonging to the limited portion; we have assumed that every point between two actual points is accessible. A point which arises by postulation may very well be actual and accessible.

Application of the theory of postulated elements. Separation. If we now allow to the postulated points, lines and planes, assuming that the theory of these has been shewn to be self-consistent, as real an existence as that which we attribute to the accessible elements, the system of real geometry which results is in accord with that considered in Chapter I, so far as the Propositions of Incidence are concerned. But care must be taken in regard to the modifications thereby introduced in the notions of a segment, and of the order of points on a line, which were fundamental in the development of the Real Geometry of the present Chapter. When postulated points are allowed an equal standing with given points, not only is it true, given a line *l* and a point *O*, that every point of this line determines a line through *O*, but, conversely, every line through *O*, in the plane *Ol*, determines a point of the line *l*; thus notions of separation, and order, for the points of the line *l*, can only retain their validity so far as similar notions can be established for the lines drawn through *O*, in this plane. Thus it appears that, without some convention as to which are inaccessible points of a line, two points of the line must be regarded as deter-

mining not one segment, but two segments, which can be distinguished from one another only by naming a point contained in
the segment referred to, in addition to the two given points. Thus,
if A, B be the given points, we may have either the segment ACB,
or the segment ADB, where C and D are two points which we speak
of as being *separated* by A and B; in this case A and B are equally
separated by C and D. An *order*, among points of the line, is now
determined by three points ; and the orders ACB, CBA, BAC are
the same, being the same as BDA, or DAB, or ABD. Thus we
can speak of four points of the line as being in order. And to
any order there is a reverse, or opposite. When points of a line,
$A, C, B, D, E, ...$, are in order, the points C and D being separated
by A and B, if $a, c, b, d, e, ...$ be, respectively, the lines joining
these points to a point, O, not on the line, then these lines are also
in order, and the lines c and d are separated by a and b. Thus,
if these lines are met by another line, respectively in the points
$A', C', B', D', E', ...$, then these points are also in order, and the
points C' and D' are separated by A' and B'. It is important to
notice, however, that *we may not speak of $A'C'B'D'$ as being the
same order as $ACBD$*. For if this were justified, the distinction
between the order $ACBD$ and its reverse, given by $CADB$, would
persist after any number of perspectivities. We have shewn however in Chapter I (p. 25 above) that, by a succession of perspectivities, we can relate the points A, C, B, D respectively to C, A, D, B.

This remark shews that the ideas involved might well be submitted
to a more minute analysis than we have entered upon.

The question at issue is intimately related with the fact that, if
all points of a line be regarded as accessible, not only does a point
of a line cease to separate the other points into two classes, but a
line in a plane ceases to separate the points of the plane, and a
plane in threefold space ceases to separate the points of the space,
and so on. For if l be a line in a plane, and A, B be two points,
not on this line, such that one of the segments AB contains a point
of the line l, the other of the segments determined by A and B will
not contain a point of this line; and a similar statement holds
when l is replaced by a plane in space, and so on. The statement
is sometimes made that, under the hypotheses adopted in Chapter I,
the plane is a *onesided* surface; but an examination and definition
of this statement would require other explanations.

Separation for four points in harmonic relation. An important result which serves well to illustrate the ideas just discussed
is that, if C, D be two points of a line which are harmonic conjugates of one another in regard to two points A, B of the line, then
these points C, D are separated by A and B.

Join the points A, B to a point R; on RC take U ; let AU, BU

meet *RB* and *RA*, respectively, in *P* and *Q*, and let *QP* meet *AB* in *D*.

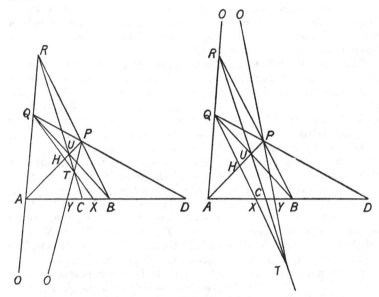

Now, on *AR* take *O*, separated from *Q* by *A* and *R*; let *OP* meet *RC* in *T* and *TQ* meet *AP* in *H*; let *TP*, *TQ* meet *AB* in *Y* and *X*; respectively.

Then, as *OAQR* are in order, so are *PAHU*, by perspective from *T*, and, thence, so are *DAXB*, by perspective of these from *Q*. Again, as *QAOR* are in order, so are *DAYB*, by perspective from *P*. Finally, as *AQRO* are in order, so are *AXCY*, by perspective from *T*.

From the orders *DAXB* and *DAYB*, wherein *D*, *A*, *B* are common and determine the order, it follows that either *DAXYB*, or *DAYXB*, are in order. Either of these contains *A*, *X*, *Y*, the former as *AXY*, the latter as *AYX*; while both *AXCY* and its reverse, *AYCX*, are in order, as we have seen. Thus either *DAXCYB*, or *DAYCXB*, are in order. In either case *C* and *D* are separated by *A* and *B* (as well as by *X* and *Y*).

This argument, which is found in F. Enriques, *Proj. Geom.* (Deutsch by Fleischer, Leipzig, 1903, p. 53), assumes that an order remains an order after perspective, and, also, as stated here, a rule for obtaining an aggregate order from two orders, specified by two sets of points containing, both, the same set of three points occurring in the same order. In regard to the possibility of a geometry (with, however, only a finite number of points) in which the point D coincides with C, the reader may consult Fano, "Sui postulati ...," *Giorn. di Mat.* xxx, 1892.

If we retain the point of view from which the present Chapter began, taking C between A and B, the point D may be only a postulated point; but the application of Peano's axiom to the theorem in question is immediate.

The impossibility of Desargues' theorem from plane Propositions of Incidence. In Chapter i we gave a proof of Desargues' theorem for two triangles in the same plane, based on a construction in three dimensions, assuming only the Propositions of Incidence. In the present Section we have given a construction for postulated points, based on Desargues' theorem for a star, which equally involves (beside Peano's axiom) the consideration of three dimensions. We now give an account of a formal proof that it is impossible to prove Desargues' theorem for the plane from the Propositions of Incidence which apply to figures in a plane, utilising for this the notions of betweenness which formed the basis of this Chapter. The reader may compare Moulton, "A simple non-Desarguesian plane geometry," *Trans. Amer. Math. Soc.*, iii, 1902; Peano, "Sui fondamenti...," *Rivista di Mat.*, iv (1894), 73; and Hilbert, *Grundlagen der Geom.*, Göttingen, 1899, § 23.

In the plane considered, let Y and U be given points, and l a

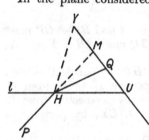

given line passing through U. Any line of the plane whose intersection with the line YU lies between Y and U gives rise to a broken line determined thus: let this intersection be M, and the intersection of the line with the line l be H; determine Q on YU so that Q and Y are harmonic conjugates of one another in regard to M and U, or say $Q = (U, M)/Y$; thus Q lies between M and U, and, therefore, between Y and U; and, conversely, if we know that Q is between Y and U, then M is also between Y and U; let the broken line consist of the portion, PH, of the line which lies on the other side of l from Y, taken with the portion HQ, these being continued indefinitely beyond P and Q, respectively. Either portion, PH, or HQ, then determines the other, and, more generally, if P' and Q' be two arbitrary points, respectively on PH and HQ, these determine the point H, and the whole broken line, conformably with the relation explained.

Let such a broken line, or a line as usually understood, be, for the moment, called a conventional line, the replacement of a usual line by a broken line being made only for lines meeting the line YU between Y and U. Then, by what has been said, the points of the plane, taken with the conventional lines, are subject to the Propositions of Incidence previously formulated in Chapter I.

But it can at once be seen that, with these conventional lines, Desargues' theorem is not always true.

For consider two triads of points, *P, Q, R*, and *P′, Q′, R′*, lying on the other side of the line *l* from *Y*, whose corresponding joins *QR* and *Q′R′*, *RP* and *R′P′*, *PQ* and *P′Q′*, meet, respectively, in points *L, M, N*, lying on the line *YU*, on the other side of *U* from *Y*; while the joins of corresponding points, *PP′, QQ′* and *RR′*, which meet in a point *T*, meet the line *YU*, respectively in points *D, E* and *F*, of which only *D* is between *Y* and *U*. Such a figure is clearly possible.

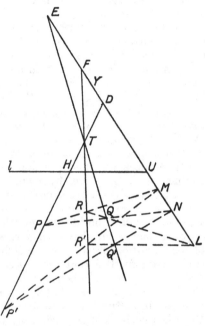

Then of the nine lines in this figure, consisting of *QR*, *RP, PQ, Q′R′, R′P′, P′Q′*, with the joins *PP′, QQ′, RR′*, there is only one which is not a conventional line by the previous definition, namely the line *PP′*; if this meet the line *l* in *H*, this would need to be modified by the replacement of *HD* by a line *HE*, to a point *E* of *YU* given by $E = (U, D)/Y$; the conventional line so obtained does not pass through the intersection, *T*, of *QQ′* and *RR′*.

Thus the Propositions of Incidence in a plane do not logically involve Desargues' theorem.

SECTION III. A DEDUCTION OF PAPPUS' THEOREM FROM THE NOTION OF ORDER AMONG THE POINTS OF A LINE, IN THE CASE OF THE REAL GEOMETRY AS AMPLIFIED BY THE ADJUNCTION OF THE POSTULATED POINTS

The abstract notion of order. We consider order in the abstract, independently of the objects to which this may be applied. These may, for example, be instants of time. To indicate this we speak of *positions*, of this order.

We conceive that the following must be granted, as necessarily inherent in the notion of such an abstract order:

(1) If two positions, say A and B, of this order, be considered, one of these, say B, is *in advance of* the other, A; and then A is *behind B*. We may denote this by $B > A$, or $A < B$; or even, as ordinary language is somewhat ambiguous, we may be allowed to say that B is to the right of A, and A to the left of B. If, then, Q is a third position which is in advance of B, it is assumed that Q is also in advance of A; and, if P is a position which is behind A, it is assumed that P is also behind B; while there may also be positions, R, which are both in advance of A and also behind B, these being said to be between A and B. In symbols, $A < B, B < Q$, together, involve $A < Q$; $A < B$, $P < A$, together, involve $P < B$; and there may be positions R for which $A < R < B$. It is assumed that every position, other than A and B, is such as Q, or P, or R. We thus build up the notion of an order of succession of any finite number of positions, determined by three of these.

(2) Between any two positions of the abstract order, there is another, and, therefore, by reapplication of the assumption, an unlimited number of other positions.

(3) Let A and B be any two positions of the order, for which, suppose, $A < B$. Consider the positions which are between A and B. Suppose that a rule has been devised, or exists, whereby everyone of these positions between A and B falls under one of two categories, (α), or (β); further, that every position falling under category (α) is seen, or known, to be behind every position falling under category (β), so that we may say that every (α) is less than every (β). Then it is in the nature of an abstract order, as we conceive it, that, there is a position of the order, say C, lying between A and B, such that every position of the order which is between A and C falls under category (α), and every position of the order which is between C and B falls under category (β). The position C itself falls under category (α) or category (β). Thus, conversely, every position (α), between A and B, if not C itself, must be between A and C; for, if it were between C and B, it would be a position (β); similarly, every position (β), if not C itself, must be between C and B.

Remarks in regard to the preceding formulation. When it is said that any set of objects are in order, it is often meant, merely, that they satisfy no. (1) of the preceding descriptions. For instance the integer numbers $1, 2, 3, \ldots$ satisfy (1); they do not satisfy (2) or (3). If we take all positive rational fractions, such as $\frac{1}{2}, \frac{2}{3}, \frac{3}{4}, \ldots$, arranged in ascending order of magnitude, they satisfy (1) and (2); they do satisfy (3). We may for instance, among these fractions, consider, as falling under category (α), those whose square is less than $\frac{1}{2}$, and, as falling under category (β), those whose square is greater than $\frac{1}{2}$; there is, then, no rational fraction occupying the position required for C in the abstract order, that is, no rational fraction whose square is $\frac{1}{2}$; and, any fraction whose square is less than $\frac{1}{2}$, has, in advance, of it, others whose squares are yet less than $\frac{1}{2}$. We may arrange these fractions, say those only which are less than 1, in another way, so that they satisfy (1), but do not satisfy (2), supposing them to follow one another in batches, of which all those in a batch have the sum of their numerator and denominator the same; the position of each fraction may then be given by one of the numbers 2, 3, 4, ... From these, and other examples, it is clear that a set of concrete objects, infinite in number, may be such as to correspond uniquely, each to one position of such an abstract order as we have explained, though yet not containing objects corresponding to every position of such order. We may give a concrete example of this, which is capable of general statement; it assumes some acquaintance with the elements of Euclidean Geometry. Take a circle, of centre O, and radius OA; let the points of the plane interior to this circle be placed in order by the following rule: through any such point, P, draw a circle, also with centre O, which intersects OA in the point A_1; let the points on the circumference of this circle be placed in order, beginning with A_1, by regarding a point Q of this circumference as being in advance of the point P of this circumference $(Q > P)$, when the positive angle A_1OQ, which is to be taken greater than or equal to zero, but *less* than four right angles, is greater than the positive angle A_1OP. Then, let all the points of the circumference of another circle of centre O, of radius OA_2 greater than OA_1 but less than OA, the first of these points being A_2, be regarded as being $>$ all the points of the circumference of the circle of radius OA_1, the points on the circumference of this second circle being, among themselves, in order by the rule above stated for the circle of radius OA_1. With such provision for any two circles of centre O within the original circle, all the points interior to this circle are arranged in order, and the above conditions, (1), and (2), are satisfied. Consider condition (3); regarding A_1 as a definite point between O and A, let the category (α) be that of all points on the circumference through A_1 and of all points on circumferences of radius less than OA_1, and let the category (β) be that of all points on circumferences of radius greater than OA_1 (but less than OA). Then every point interior to the circle of radius OA falls under one of these categories, and every point of category (α) is $<$ every point of category (β). But there is no point, C, such that every point $< C$ belongs to the category (α), and every point $> C$ belongs to the category (β). If we take any point, Q, of the circumference through A_1, there are always points,

Q', of this circumference, which are $> Q$; and then there are points, Q'', of this circumference, which are $>$ any such points Q', and so on continually; the process will not have an end unless either the point A_1 is allowed as $>$ all the various points Q, Q', Q'', ..., which, however, is not the case, since $A_1 < Q < Q' < Q''$ ..., or some point is allowed $>$ all the various points Q, Q', Q'' ... which is a point of a circumference of radius greater than OA_1; and this, again, does not provide the point C in question, since all such points are of category (β). Thus, while the construction furnishes points arranged in such an abstract order as we have described, it does not give points corresponding to every position of such an order.

A further remark may be made: The notion of an abstract order furnishes no means for saying that the positions of the order between two positions A and B form an aggregate of the same *length*, or *magnitude*, or *power*, as the aggregate of positions between two other positions C and D. Such a notion of *congruence* must be supplied, if at all, by additional conditions. And only when this is supplied can there be question of the fulfilment of what is known as the axiom of Archimedes: the axiom namely, for an aggregate of elements arranged in order wherein the notion of equality of intervals between pairs of elements has been established, which states that, if A, B be two elements, and P_1, P_2, ... be elements, such that $A < B$ and $A < P_1 < P_2 < ...$, and also such that the intervals AP_1, P_1P_2, P_2P_3, ... are all *equal*, then there is a number n for which $B < P_n$. We may, as before, be allowed, in this case a very simple and concrete illustration, for the purpose of helping to fix the ideas; though, like a diagram used to assist the conception of a geometrical figure, the illustration may suggest implications which are not intended: In the case of time we measure intervals by observing the succession of natural phenomena, as for instance by heart-beats, or by a clock. But if all the clocks went twice as fast between what we now call 12 o'clock and 12.30 o'clock, as they now do, so that they indicated 1 o'clock at the time we now call 12.30, and went two-thirds as fast as they now do between the times we now call 12.30 and 2 o'clock, so that they marked 2 o'clock as they now do, we should, if we had agreed to regard as standard time that given by the clocks, reckon the interval between our present times 12 and 12.30 as being equal to the interval between 12.30 and 2 Though this would lead to modifications in our description of the behaviour of natural bodies, as, for instance, the contractions of the heart, it would not affect our notion of the succession of time, or of the order of events. The questions suggested in this connexion, recently also in vogue as arising in discussions of the Principle of Relativity in Physical Science, are inherent in the consideration of the Foundations of Geometry. The Theory of Proportion in Euclid's Fifth Book is evidence of their antiquity. In modern literature, the analysis leading to condition (3) above, relating to the existence of, a dividing position there called C, was suggested by R. Dedekind, *Stetigkeit und irrationale Zahlen*, Braunschweig, 1892, p. 11. For the relation of number and the measurement of quantity an article by O. Hölder, *Die Axiome der Quantität*, Leipzig, Berichte, LIII, 1901, may be consulted; and for this, and other fundamental questions, see F. Enriques, *Encyklopädie der Math. Wiss.*, III, 1. 1, *Prinzipien der Geometrie*, where many references are given.

A theorem in regard to corresponding positions of an abstract order. Let us now suppose that we have some means of establishing a correspondence between two positions, P and Q, of such an abstract order, so that to any position P there corresponds a position Q, and conversely; this being such that the positions Q', Q'' corresponding to two positions P', P'', for which $P'' > P'$, are also such that $Q'' > Q'$. Let us suppose, further, that there are two positions of P, say A and B, for each of which the corresponding position of Q coincides with that of P; and suppose that the position of Q does not coincide with that of P for every position of P between A and B. When P has a position between A and B, the position of Q, if not coincident with P, relatively to P, may agree with or differ from that of B relatively to A; it will be sufficient to, and we shall, suppose $A < B$, and the two possibilities are, then, $Q > P$ and $Q < P$, the former being that when Q is in advance of P, and the latter that in which Q is behind P. It may be that Q, which coincides with P when P is at A, and coincides with P when P is at B, is, otherwise, for every position of P between A and B, in advance of P; or is, for every position of P between A and B, behind P. In either case we may say that Q *preserves the same relation to P* when P is anywhere within AB. More generally, if Q coincides with P when P is at each of two positions, P_1 and P_2, between A and B, and Q preserves the same relation to P when P is anywhere within P_1P_2, being either in advance of P for every position of P between P_1 and P_2, or behind P for every position of P between P_1 and P_2, it will be convenient to speak of P_1P_2 as a *complete interval*. We desire, then, to prove, assuming, as we have said, that Q does not coincide with P for every position of P between A and B, that, if AB is not a complete interval, it contains complete intervals, the end positions of these, whereat Q coincides with P, being definite. For this, consider a position, P_0, of P, between A and B, for which the corresponding position, Q_0, of Q, is not coincident with P_0. For definiteness suppose that Q_0 is in advance of P_0; a similar argument is applicable when Q_0 is behind P_0; it may be that, as P takes positions between P_0 and B, Q preserves that relation to P (of being in advance of P) which it has when P is at P_0, coinciding with P only when P is at B. In any case we prove that there is a definite position, V, between P_0 and B or identical with B, such that Q is in advance of P for every position of P between P_0 and V, but coincides with P when P is at V. And a similar argument will establish the existence of a definite position, U, between A and P_0 or identical with A, such that Q is in advance of P for every position of P between U and P_0, but coincides with P when P is at U. The segment UV will then be complete, in the sense adopted. For greater definiteness, let any position P between P_0 and B be said to be *like* P_0, when the corresponding position of Q is in advance of that position of P, and be said to be unlike P_0, when the corresponding position of Q either coincides with that position of P or is behind it. In particular B is unlike P_0. We may then place every position, F, between P_0 and B, under one of two categories, (α) or (β), agreeing that F is of category (α) when every position between P_0 and F, and F itself, is like P_0, but that F is of category (β) when there is one or more position between P_0 and F, or at F, which is unlike P_0. There would be no meaning in saying that P_0 is of either category, but B may be regarded as being of category (β), since Q there coincides with P. A position of category (β) cannot coincide with a position of category (α); nor can any position of category (β) lie between P_0 and a position of category (α). In other words every position of category (α) is behind every position of category (β). Thus, recalling the condition (3) attached to the notion of an abstract order, there exists a position, V, between P_0 and B or at B, such that all positions of P in P_0B which are behind V are of category (α), and all positions in advance of V of category (β). This is equivalent to saying that every position, X, of P_0B, which is behind V, is like P_0, and every position of P_0B which is in advance of V, is either unlike P_0 or has positions unlike P_0 lying behind it; thus every position, Y, of P_0B, in advance of V, has positions unlike P_0 lying behind it; for every position between V and Y, being in advance of V, is either itself unlike P_0 or has such positions lying behind it.

Hence, any position, X, of P_0B, which is behind V, is such that, when P is at X, the position Q is in advance of X; and every position Y, of P_0B, which is in advance of V, is such that, between V and Y, there is a position of P for which the corresponding Q is behind P, or a position P for which the corresponding Q coincides with it.

Now introduce the hypothesis we have made, that, if Q_1, Q_2 be the positions of Q corresponding to two positions P_1, P_2 of P for which P_2 is in advance of P_1, then Q_2 is in advance of Q_1. Then we can shew that, when P is at V, so also is Q. For if, when P is at V, the corresponding Q were at Q', behind V, and we take a position of P between Q' and V, and, therefore, behind V, the corresponding Q would, by the property of V, be in advance of P, and, therefore, in advance of Q', though the P to which this corresponds is not in advance of the P to which Q' corresponds, which is at V. Or, again, if, when P is at V, the corresponding Q were in advance of V, say at Q'', and we take a position of P between V and Q'', and, therefore, in advance of V, the corresponding Q would, by the property of V, either be at this P or behind it, and in either case behind Q'', though the P to which this corresponds is not behind the P to which Q'' corresponds, this being at V.

Thus, it is proved that, when P is at V, the corresponding Q coincides with it, there being between P and V no position of P for which Q is not in the same relation to P as when P is at P_0. And, in particular, V may be B.

The proof of the existence of the position U is similar; but, as it differs in detail, it may also be given. As before, suppose that $Q_0 > P_0$, and let positions of P, between A and P_0, for which Q is in advance of P, be said to be like P_0, positions for which Q coincides with, or is behind P, being said to be unlike P_0; so that A is unlike P_0. Every position, F, between A and P_0 or at A, is either, (α), such that every position between F and P_0, and F itself, is like P_0, or, (β), such that there is one or more positions between F and P_0, or at F, which is unlike P_0. Then every (α) is in advance of every (β); and, by the condition (3) for an abstract order, there exists a position, U, between A and P_0, or at A, separating positions (α) from positions (β). Thus, every position, X, between U and P_0, is like P_0, and every position between A and U is either unlike P_0 or has such positions lying in advance of it; from which, every position, Y, between A and U, has positions unlike P_0 lying in advance of it. Or, when P is at X, the corresponding position Q is in advance of it, while between Y and U there are positions of P for which Q, if not coincident with P, is behind P. And, as before, we can shew that when P is at U, so is Q. For, if the position U for Q arose from a position, P', of P, lying between U and P_0, a position Q, between U and P', would, by the property of U, correspond to a position P behind Q, and, therefore, behind P', and we should have the results $P' > P$, $U < Q$, which, by hypothesis, are inconsistent; while, if the position U for Q corresponded to a position P'' between A and U, then a position Q between P'' and U would correspond to a position which, if not coincident with Q, is in advance of Q, and is, therefore, in advance of P''; then we should have the inconsistent results $P'' < P$ and $U > Q$.

The existence of the complete segment is thus clear.

Application of the theory of an abstract order to the points of a line. Deduction of Pappus' theorem. We proceed now to apply the preceding work to the consideration of the points of a line. In doing this, we recognise only the real points given by the preliminary propositions and constructions of the earlier part of this Chapter, with the postulated points to which these have led. It is then clear that such points are in such an order as satisfies the condition (1) and (2), above given, for an abstract order. Let us now *assume that the points of a line arising by the preliminary propositions and constructions, together with the postulated points, also satisfy the condition* (3). Thus, if the points of any segment, arranged in order, be all of one of two categories, (α), (β), of which every (α) precedes every (β), then there is a real point of the segment separating the points of category (α) from those of category (β), except that this point itself is of one of the two categories.

To the properties of the aggregate of the points of a line assured by this assumption, the constructions of geometry, in the space in which the line lies, then add two further properties: They enable us to establish such a correspondence between two points of the line as has been supposed in the immediately preceding discussion of an abstract order; and they enable us, given a segment, *AB*, of the line, and a point, *C*, not lying in this segment, to obtain a point, *C'*, which does lie in this segment, namely, the harmonic conjugate of *C* in regard to *A* and *B*. For we have shewn that *C* and *C'* are separated by *A* and *B* (above, p. 119).

Now consider four points, *P, Q, R, S*, upon a line, and suppose that these are in order. We have explained (above, p. 118, in Section II of Chapter II) that, if four points upon one line are in perspective, respectively, with four points upon another line, and if one set of four be taken in order, then the points of the other line, respectively corresponding to them, are also in order. Hence, if, through an arbitrary line *l* not meeting the line *PQRS*, there be drawn planes, each containing one of the four points *P, Q, R, S*, these planes, like *P, Q, R, S*, will be in order; and if these planes be, respectively, met by another line in points *P', Q', R', S'*, then *P', Q', R', S'* will also be in order.

This being premised, let *a, b, c, d* be four lines, of which no two intersect. We proceed to deduce that, if these four lines are all met by three others, then they are met by an infinite number of others, one through every point of any one of the four given lines. It was previously shewn (above, p. 49) that this theorem is equivalent, in virtue of the Propositions of Incidence, with Pappus' theorem. Let the points in which the three assumed transversals meet one of the four given lines, *d*, be, respectively, *A, B* and *C*. Of the two

segments *AB*, consider that one which does not contain the point *C*.
Let *P* be any point of the line *d*; from *P* draw the transversal to
meet the lines *a* and *c*, its point of
meeting with *c* being *O*; from *O* draw
the transversal to meet the lines *b*
and *d*, its intersection with *d* being *Q*.
Then, to any position of *P* on the
line *d* corresponds a definite position
of *Q*, and, conversely, to every *Q* a
definite *P*. The various positions of
P correspond, each to one of an axial
pencil of planes, having the line *a*
for axis, the points *P* being the in-

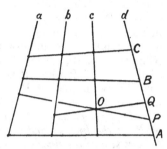

tersections of the planes of this pencil with the line *d*; three of
these planes, in particular, are those containing, respectively, the
points *A*, *B* and *C*. The planes of this pencil, respectively, are
also intersected in points by the line *c*, the point *O* of this being
that which corresponds to *P*. But, again, the various positions of
O, give rise, with the line *b* for axis, to an axial pencil of planes,
of which every plane has also a point on the line *d*, the point *Q*
being that so arising from *O*. Therefore, four positions of *P* which
are in order, give rise to four positions of *O* which are in order,
and these, in their turn, to four positions of *Q* which are in order.
In particular, *Q* coincides with *P* when *P* is at *A*, or at *B*, or at *C*.
Thus, if P_1, P_2 be two positions of *P* such that *A*, P_1, P_2, *B*, *C* are
in order, and if Q_1, Q_2 be the positions of *Q* respectively corresponding to P_1 and P_2, then, also, *A*, Q_1, Q_2, *B*, *C* are in order.
We may regard the points *ABC* as defining, for the points of the
line *d*, such an order as was contemplated in the discussion of an
abstract order above, which is also such as was considered at the
beginning of the present chapter (p. 96, above), where the inaccessible points of a line in fact play the part here played by the
point *C*. We may then say, limiting the position of *P*, now, to
the segment *AB* which does not contain *C*, that: *To any position
of P between A and B there corresponds a definite position of Q, also
between A and B, and conversely; and, if Q_1, Q_2 be the positions of
Q corresponding to the positions P_1, P_2, of P, of which P_2 is in advance of P_1, then, also, Q_2 is in advance of Q_1.*

It therefore follows, from the preceding theory, that, unless, for
all positions of *P* between *A* and *B*, the position of *Q* is coincident
therewith, there can be found, between *A* and *B*, two positions,
U and *V*, which may coincide, respectively, with *A* and with *B*, such
that, when *P* is at either *U* or *V*, then *Q* coincides with *P*, while,
for a position of *P* between *U* and *V*, the position of *Q* preserves
the same relation to that of *P*, and never coincides with it. This

last alternative is, however, untenable: For, if we take the harmonic conjugate of the point C in regard to U and V, we obtain a point, C', lying between A and B, as has been proved (p. 119, above). By the construction given, four positions of P which are in harmonic relation, give rise to four positions of Q also in harmonic relation, as follows from the properties of this relation (Chapter I, p. 13, above). Therefore, as the positions of Q coincide with those of P when P is at U, or V, or C, the position of Q coincides with that of P when P is at C'. The remaining alternative must then be taken: *The positions of P and Q are coincident for all positions of P between A and B; and hence for all positions of P on the line d; and the transversal drawn from any point of d to meet two of the lines a, b, c, also meets the third.*

As has been said, it follows from the theory given in Section II of Chapter I (p. 49, above), that this amounts to a deduction of Pappus' theorem, under the assumptions made above as to the points that are possible upon a line. It proves also, with the same assumptions, that if, of two related ranges on two lines, possibly coincident, there be given three points, A', B', C', of one range, to correspond, respectively, to three points, A, B, C, of the other, then the point, D', of the former range, which corresponds to an arbitrary point D, of the latter, has a unique position (see pp. 45 ff., above). In particular, if, of two related ranges on the same line, there be three common points, each self-corresponding, then the ranges are identical.

Similar statements then follow in regard to flat pencils, of lines in a plane all passing through the same point, related to one another; and in regard to axial pencils, of planes in space all passing through the same line, related to one another.

Alternative deduction of related ranges, based on a harmonic net. In what has preceded in this volume, two ranges have been related to one another by a sequence of perspectivities. In a perspectivity four points of a line which are in harmonic relation lead to four points of another line which are also in harmonic relation. Conversely, suppose that we can set up a correspondence between the points of two lines, of such character that to any point of either line there belongs a definite point of the other, in such a way that any four points of either line which are in harmonic relation, say A, B, C, D, of which B, D are harmonic conjugates of one another in regard to A and C, correspond to four points of the other also in harmonic relation, say, respectively, A', B', C', D', of which B', D' are harmonic conjugates of one another in regard to A' and C'; suppose further that the order $A'B'C'$ is the same as, or different from, a definite order previously established on the second line, according as the order ABC is the same as, or different from, a definite order previously established on the first line; then it may well appear probable that the two ranges are related, in the sense previously employed.

We do not enter fully into the analysis of this suggestion. But there is one result which would be necessary for this purpose, to which we should refer; it has an interest and importance by itself, and its deduction is by reasoning cognate to that immediately preceding. Suppose that, upon any line, are taken three points, A, B, and C; and, then, the harmonic conjugate of each of these in regard to the other two, and, then, further, the harmonic conjugates of each of the new points in regard to every other two, and so on continually. The points so obtained will have the property that, in each of the two segments determined by any two of the points, there are other points of the set; for beside these two points there are certainly others, and one such, lying in one of these segments, necessitates one, the harmonic conjugate of the other, lying in the other segment. This property, however, by itself, does not secure that, if two *arbitrary* points be taken on the line, not assumed to be points of the constructed set, there is a point of the set lying in a specified one of the two segments determined by these arbitrary points, though, if not, such points are in the other segment. In virtue of the special character of the harmonic relation it is however the case that this is so, and it is this fact which we now prove.

The definition of related ranges here referred to, founded on the condition that four harmonic points always correspond to four harmonic points, is that which was adopted by von Staudt (*Geometrie der Lage*, 1847, § 9, p. 49), who assumed without proof the result we have stated—that a point of a harmonic net of points constructed as above is found in every arbitrary segment. This omission was remarked by F. Klein, who gives, *Math. Annal.* VII, 1874, *Nachtrag*, p. 535, the following demonstration, as due to Zeuthen, stating it to be equivalent to another simultaneously communicated by Lüroth.

Two assumptions are made, into a detailed analysis of which we need not enter:

(1) If, in regard to two points A, B, the harmonic conjugates of two points C, C' be respectively D and D', then the orders, $CC'B$ and $DD'B$ are different, as also are the orders ACC' and ADD'. In other words, if $ACC'B$ be in order, then $ABDD'$ are not in order; but if $ADD'B$ are in order, then $ACC'B$ are not in order.

For let A, C, B be in perspective, from a point O, with A, R, P, respectively, and A, R, P be in perspective, from B, with A, Q, O, respectively, and, lastly, A, Q, O be in perspective, from P, with A, D, B, respectively. If then C' be in the segment CB which does not contain A, it gives rise, by perspective from O, to a point, R', in the segment RP which does not contain A; and R', by perspective from B, gives rise to a point, Q', in the segment QO which does not contain A; and, then, Q', by perspective from P, gives rise to

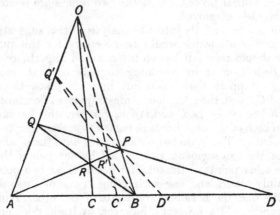

the point, D', in the segment DB which does not contain A; and D' is the harmonic conjugate of C' in regard to A and B.

The assumption may be expressed by saying that points which are harmonic conjugates of one another, in regard to two fixed points, move in opposite directions.

(2) If of the four points $ACBD$, in harmonic relation, we modify,

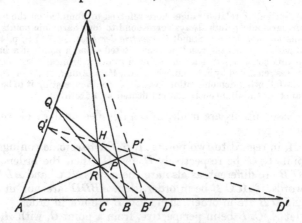

not C and, consequently, D, but, keeping C the same, we modify B, and, consequently, D, then, when $ACBB'$ are in order, so are $ABDD'$, and conversely; the orders CBB' and CDD' are the same, as are ABB' and ADD'.

For, with the same construction for D, from C, as in (1), let a point, B', of AB, be joined to R and O by lines meeting AQ and

AP respectively in Q' and P', and $Q'P'$ meet AB in D', so that D' is the harmonic conjugate of C in regard to A and B'; then, if B' be in the segment AB which does not contain C, it follows, by perspective from O, that P' is in the segment AP which does not contain R, and, by perspective from R, that Q' is in the segment AQ which does not contain O. Hence, remarking that $OR, PQ, P'Q'$ meet in a point, H, as follows from the triads P, P', O and Q, Q', R, it is seen that D' is in the segment DA which does not contain B.

In (1) two separated points, of four in harmonic relation, were kept unaltered; in (2) two contiguous points are to be kept unaltered. Then, we may say, the other two, contiguous, points move in the same direction.

These conditions being understood, consider such a set of points on a line as was described, containing the harmonic conjugate of every one of them in regard to every other two. We may speak of such a set as a harmonic net; and may associate with it the name of Moebius. Suppose, if possible, there are two points, F, G, of the line, such that there are no points of the harmonic net on one of the two segments determined by F and G. Points of this segment may be spoken of as lying between F and G; and, by taking a further point of the line to define, with F and G, an order, we may, without loss of generality, speak of G as being in advance of F. We first extend this segment FG as much as possible, seeking to replace F by a point X behind F, and replace G by a point Y in advance of G. Precisely, we shew that there are two definite points, X, Y, upon the line, of which X either coincides with F or is behind it, and Y either coincides with G or is in advance of it, so that $XFGY$ are in order, having the properties, (a), that there is no point of the harmonic net within the segment XY, which contains F and G, (b), that either X is a point of the harmonic net, or there are points of the harmonic net lying between X and any *arbitrary* point taken behind X, (c), that either Y is a point of the harmonic net, or there are points of the harmonic net lying between Y and any arbitrary point taken in advance of Y.

To prove the existence of the point Y, we remark, first, that if G be a point of the harmonic net we may take Y at G. Suppose, that G is not a point of the harmonic net; there must, then, be points of the harmonic net in advance of G, or behind F, since otherwise, there being none between F and G, there would be none anywhere. In our conventions of order, however, there is, finally, no distinction between points said to be in advance of G, and points said to be behind F. We shall therefore say that there is a point, N, of the harmonic net which is in advance of G. Consider now the segment GN, containing points in advance of G, but not containing F. We apply the formal argument introduced in our discussion of an abstract order, to shew that there is such a point as Y within this segment. The points of the segment GN, other than G, are of one of two categories: (α), Points, R, such that there is no point of the harmonic net between G and R, or at R, the word *between* referring to that segment GR which does not contain F; (β), Points, S, such that there is a point of the harmonic net between G and S, or at S. It is clear that every point (α) is behind every point (β). Therefore, by the condition named (3) in the discussion of an abstract order, there is a point, Y, between G and N, or at N, such that all points of the segment GN which are behind Y are points (α), and all points of the segment YN are points (β), the point Y, itself, being of one of these categories. Every such point (β) of the segment YN, say S, is then either, itself, a point of the harmonic net, or has such points, lying in GN, behind it. If behind S there be no such harmonic points, so that, in particular, Y is not a harmonic point, and if, further, there are no harmonic points between Y and S, in which case S itself is a harmonic point, then there are points (α) in the segment YS, which is contrary to the definition of Y. Wherefore, every point, S, of category (β), in advance of Y, has points of the harmonic net behind it, and these, by the construction, are not behind Y. Thus, either Y is a harmonic point, or such harmonic points lie between Y and R, whatever point of the segment YN the point R may be. Thus Y is such a point as was to be shewn to exist. The existence of X can be shewn in a similar manner.

Having thus deduced, from the hypothesis of the existence of the segment FG containing no points of the harmonic net, the existence of X and Y, with no harmonic points between them, we proceed, with use of the precise definitions of X and Y, to give a construction for finding a point of the harmonic net which does actually lie between X and Y. We shall then be able, finally, to conclude that such a segment as FG is not possible.

For this purpose, suppose, first, that neither X nor Y is a point of the harmonic net. And, for the sake of clearness, denote points of the harmonic net which arise in the construction by A_1, B_1, C_1, D_1, points not assumed to belong to the harmonic net being denoted by letters without suffix. Then make the following construction:

(1) Take any point, A_1, of the harmonic net, not in the segment XY containing the segment FG first considered, the point A_1 being, say, behind X. From A_1 find the harmonic conjugate in regard to X and Y, say H, represented by

$$H = (X, Y)/A_1,$$

so that H is within the segment XY.

(2) Take also the point J given by

$$J = (A_1, Y)/X,$$

so that J is in advance of Y.

(3) Remembering the property of the point Y, take a point, B_1, of the harmonic net, in advance of Y but behind J, and then a point, T, in advance of Y but behind B_1, and thence find the point, U, such that

$$U = (A_1, H)/T.$$

This shews that U is in advance of A_1 but behind H; while, as we also have

$$X = (A_1, H)/Y,$$

we can infer, from the assumption (1) above, that the order $A_1 X U$ is the opposite of the order $A_1 Y T$, namely that U is behind X, and it is in advance of A_1.

(4) By the property of the point X, there is a point of the harmonic net in advance of U and behind X; take such a point, C_1. From this find K so that

$$K = (A_1, H)/C_1;$$

then, as UC_1XY are in order, and A_1, U, H, T and A_1, X, H, Y, as well as A_1, C_1, H, K, are all sets in harmonic relation, it follows, also from assumption (1) above, that $TKYX$ are in order, or that K is in advance of Y and behind T.

(5) Next take L so that

$$L = (C_1, J)/A_1.$$

Then, comparing the facts that A_1, J, L, C_1 and A_1, J, Y, X are both sets in harmonic relation, we infer, from assumption (2) above, that A_1, L, Y are in the same order as A_1, C_1, X, or that L is behind Y. While, comparing the facts that A_1, C_1, H, K and A_1, C_1, L, J are also both sets in harmonic relation, we similarly infer that A_1, H, L are in the same order as A_1, K, J, or that L is in advance of H.

(6) Lastly take D_1, belonging to the harmonic net, such that

$$D_1 = (C_1, B_1)/A_1.$$

Then, from the harmonic sets, A_1, C_1, D_1, B_1 and A_1, C_1, L, J, we similarly infer that C_1, D_1, L are in the same order as C_1, B_1, J, or, that D_1 is behind L. And, from the harmonic sets, A_1, C_1, H, K and A_1, C_1, D_1, B_1, we infer that the order $A_1 K B_1$ is the same as $A_1 H D_1$, so that D_1 is in advance of H.

We may then say that D_1 is between H and L. But, similarly, H is between X and Y, and L is between H and Y. Thus D_1 is between X and Y; and is a point of the harmonic net. Wherefore, in the case when neither X nor Y is a point of the harmonic net, we are able to infer that the hypothesis of the segment FG is untenable.

Next suppose that X is a point of the harmonic net, but, as before, that Y is not such a point. Then the point C_1, which was any point of the harmonic net between U and X, can be taken, instead, to be at X. Hence, by comparison of (3) and (4), the point K will be at Y; and, by comparison of (2) and (5), the point L will be at Y. But D_1 will still be between X and Y; and the same conclusion will follow.

If Y, but not X, be a point of the harmonic net, we can interchange the parts played by X and Y in the preceding case, with the same inference.

While, finally, if both X and Y belong to the harmonic net, the harmonic conjugate of A_1 in regard to X and Y is between X and Y, and, in this case, is a point of the harmonic net.

The conclusion, then, is, that every segment of the line contains one, and, therefore, an infinite number of points of the harmonic net.

Consequences of the equable distribution of points of the harmonic net. It follows from the preceding theorem that any point, P, of a line upon which a harmonic net has been constructed, may be regarded as lying in advance of an indefinitely continued sequence of points of the harmonic net, say $A_1, A_2, A_3, ...$, where $A_1A_2A_3...P$ are in order, and also as being behind another such sequence, $B_1, B_2, B_3, ...$, where $P...B_3B_2B_1$ are in order. We have only to take two arbitrary points, A, B, upon the line, such that APB are in order; there is, then, a harmonic point, A_1, between A and P, and a harmonic point, B_1, between P and B. There is, also, a harmonic point, A_2, between A_1 and P, and another, B_2, between P and B_1; and so on, indefinitely. Conversely, if we have an indefinitely continued sequence of points of the harmonic net, in order, say $A_1, A_2, A_3, ...$, and, also, another indefinitely continued sequence of harmonic points ..., B_3, B_2, B_1, in the same order, and if every point of the former set lies behind every point of the latter, there is, by the condition (3) introduced in the discussion of an abstract order, a point P which has the specified relation to the two given sets of harmonic points. This point P may be, or may not be, a point of the harmonic net.

Another result, involving similar ideas, should be referred to here. If O, E, U be any three points of a line, and we carry out in succession the operations of finding the harmonic conjugate, E_1, of O in regard to E and U; then the harmonic conjugate of E in regard to E_1 and U, say E_2; then the harmonic conjugate of E_1 in regard to E_2 and U, say E_3, and so on, indefinitely, the points $OEE_1E_2E_3...U$ will be in order. And, if T be any point of that

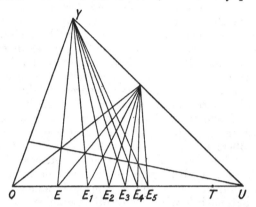

segment EU which does not contain O, there will be found points of the set E_1, E_2, E_3, ..., in infinite number, in that segment TU which does not contain E, *and this wherever T be taken.* This result is like one referred to previously (p. 124, above), applicable to measurable quantities, known as Archimedes' axiom, which would state that, if EE_1, E_1E_2, E_2E_3, ... were *equal lengths,* and T arbitrary, there is an n such that OE_n is greater than OT. But, as is clear, the fundamental ideas are different: the result here stated is a theorem of 'betweenness,' when all the points of the line are supposed accessible; Archimedes' axiom is a theorem of that which is beyond what is given.

Introduction of symbols, in particular of the iterative symbols. We have, in this chapter, avoided reference, so far, to the algebraic symbolism described in Chapter I, and to the numbers of ordinary arithmetic. But the effect of the hypothesis introduced, for the points of a line, to obtain a deduction of Pappus' theorem, that the real points existing on a line are such, and only such, that condition (3) for an abstract order (p. 122, above) is satisfied, is to set up a correspondence between the points of a line and the numbers of arithmetic. In this correspondence, the numbers are used only to specify order of succession, not magnitude, and there is one postulated number, denoted by ∞, which is to be regarded as coming not only after all positive numbers but also before all negative numbers; this being understood, the numbers associated with three arbitrary points of the line may be taken arbitrarily, and the correspondence is then definite.

Taking three arbitrary points of the line O, E, U, and supposing the symbols belonging to these to be such that $E = O + U$, as was explained in Chapter I, we have obtained an interpretation, by a point of the line, of the symbols $nO + U, O + mU, nO + mU, nO - mU$, wherein m, n are iterative symbols; and it is easy to establish, by induction, from the assumptions of the present chapter, that, if the positive numbers of arithmetic, $\bar{n}, \bar{m}, \bar{n}', \bar{m}'$, with which the symbols n, m, n', m' are formed, be such that $\bar{m}'/\bar{n}'$ is greater than $\bar{m}/\bar{n}$, then the points of which the symbols are, respectively, $O, nO + mU$, $n'O + m'U, U$, are in order; with a similar conclusion for points of symbol $nO - mU$. By this, a correspondence is established between the points of a harmonic net upon the line, built up from the points, O, E, U and the rational numbers of arithmetic, taken in order in both cases, the points O, E, U being associated respectively with the numbers $0, 1, \infty$. Then, by the proposition of the equable distribution of the points of the harmonic net, proved above, the points of the line not belonging to the harmonic net are made to correspond, uniquely, each to an irrational number. It will be seen, immediately, below, that any other numbers than $0, 1, \infty$, might have been attached to the points O, E, U.

It is not, however, thereby asserted, as has several times been

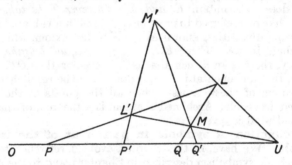

remarked, that, if two points, P, Q, of the line be thus associated with two symbols p, q, respectively, and two other points, P', Q', with two symbols p', q', and if $q' - p' = q - p$, then the *length* $P'Q'$ is equal to the length PQ. Such an assertion would be meaningless in the absence of a definition of length. As we have seen, the points $P, Q, \dots$ being $O + pU, O + qU, \dots$, and therefore, the points P', Q, Q' being, respectively, $P + (p' - p) U, P + (q - p) U$, $P + (q' - p) U$, the equation $q' - p' = q - p$, or $q' - q = p' - p$, expresses that, if arbitrary lines be drawn through P, P', Q, respectively, in one plane containing the line, that through P meeting those through P' and Q respectively in L' and L, and UL, UL' meet the lines through P' and Q respectively in M' and M, then the line $M'M$ meets the original line in the point Q'. Evidently, when P, P' and Q are assigned, the position of Q' depends on the position of U.

Thus the following question arises: If we take three arbitrary points of the line, O, $O + U$, U, and thereby associate every point of the line, of symbol $O + xU$, with a number, which we shall denote by x; and also take three other arbitrary points, denoting these by O', $O' + U'$, U', and thereby associate every point, $O' + x'U'$, with a number x'; what is the relation of the numbers x', x, when the symbols $O + xU$, $O' + x'U'$ refer to the same point? By denoting an arbitrary point by $O' + U'$, we have fixed the multiplier of the symbol U'; we have, therefore, syzygies

$$O' = mO + aU, \quad U' = bO + cU,$$

where m is arbitrary, and may be replaced by 1 unless O' coincides with U, but $m^{-1}a$, b, and c, are definite. Thence

$$O' + x'U' = mO + aU + x'(bO + cU), \ = (m + x'b)O + (a + x'c)U;$$

identifying this with $O + xU$, we have a relation connecting x and x'. We saw, in Chapter I, that, without assumption of Pappus' theorem, the iterative symbols, corresponding to the rational numbers of Arithmetic, are commutative in multiplication; the hypotheses of the present chapter involve Pappus' theorem, and therefore that the symbols of all the real points of the line recognised by these hypotheses, are also commutative in multiplication. We may, therefore, write

$$x = \frac{a + cx'}{m + bx'}, \quad x' = \frac{mx - a}{c - bx}.$$

In other words, the use of the fundamental points O', $O' + U'$, U', instead of O, $O + U$, U, leads to the substitution of x by a general linear (fractional) function of x, involving three coefficients.

Consider, now, two related ranges upon the line. By what we have seen, two such ranges are obtained in the most general way by taking three points O, E, U, of one range, to correspond, respectively, to three points, O', E', U', of the other. In particular, any set of four points, of the first range, which are in harmonic relation, gives rise to four points of the second range in similar relation. Thus, every point of the harmonic net constructed from O, E, U, corresponds to a point, constructed, with precisely similar steps, from O', E', U', belonging to the harmonic net built up from these. By using the symbols U, U' with proper multipliers, we can suppose the symbols E, E' to be such that $E = O + U$, and $E' = O' + U'$. Then the point $P = mO + nU$ corresponds to the point $P' = mO' + nU'$, where m, n are iterative symbols; and, hence, $P = O + xU$ corresponds to $P' = O' + xU'$, where x is any symbol for a real point of the line. When $O' = mO + aU$, $U' = bO + cU$, as above, this last is $P' = (m + xb) O + (a + xc) U$. Thus: *In order that a range of points $P, = O + xU$, for varying x, should be related to a range of points $P', = O + yU$, the symbols x, y must be connected by an equation of the form $y = (cx + a)/(bx + m)$*. By proper choice of c, a, b, m, it is possible, in this formula, to make three arbitrary points of the first range correspond, respectively, to three arbitrary points of the second; the formula appears thus to cover the case of every two possible related ranges.

Ex. 1. In the construction above given to prove the equable distribution of the points of a harmonic net, prove that the symbols of the various points may be supposed to be of the following forms, respectively,

$$X = A_1 + Y, \quad H = A_1 + 2Y, \quad J = A_1 - Y, \quad B_1 = A_1 - bY, \quad T = A_1 - tY,$$
$$U = (t + 1) A_1 + tY, \quad C_1 = (c+1) A_1 + cY, \quad K = A_1 - cY, \quad L = A_1 + 2cY$$

leading to
$$D_1 = (bc - c + b) A_1 + 2bcY$$
$$= (b-1) c^2 H + (c - b) L,$$

the numbers of arithmetic belonging to the symbols c, t, b, 1, being such that
$$c > t > b > 1,$$

shewing that, then, the points are in the order required in the demonstration.

Ex. 2. In the Ex. 4, p. 84, of Section III of Chapter I, shew that it is impossible, consistently with the assumptions, in regard to order, adopted in this chapter, for the point $Q = O + a^2 U$ to coincide with the point $O - U$, obtained as $(O, U)/E$. Give constructions for which $O + a^3 U$ coincides with $O + U$, or $O - U$.

Ex. 3. If the points $O + x'U$, $O + xU$, be harmonic conjugates in regard to the points $O + aU$, $O + bU$, prove that
$$x' = [x (a + b) - 2ab] \div [2x - a - b].$$

Using the same letters for the associated numbers, deduce that
$$\frac{\partial x'}{\partial x} = - \frac{(a-b)^2}{(2x - a - b)^2}, \quad \frac{\partial x'}{\partial b} = 2 \frac{(x-a)^2}{(2x - a - b)^2};$$

these are in accordance with the two assumptions made for the proof of the equable distribution of the points of a harmonic net (above, p. 132).

CHAPTER III

ABSTRACT GEOMETRY, RESUMED

RELATED SPACES; JUSTIFICATION OF THE SYMBOLS. GEO-
METRICAL ASSUMPTION FOR IMAGINARY ELEMENTS; RE-
PLACEMENT OF IMAGINARY ELEMENTS BY SERIES OF REAL
ELEMENTS

General discussion of the problems of this chapter. The
preceding chapter has shewn how, by the recognition of the order
of the real points of a line, and by adjunction of postulated points
defined by constructions in a limited accessible space, we can build
up a geometry in which, with a definite assumption as to the points
existing upon a line, Pappus' theorem can be proved. The Real
Geometry so arising may then be regarded as a particular case of
the Abstract Geometry considered in Chapter i; it was seen that
the algebraic symbols which arise in this Real Geometry are precisely similar, in their mutual relations, to the numbers of real
arithmetic.

The present chapter is a continuation of Chapter i. It was there
shewn that upon the Propositions of Incidence, and the assumption
of Pappus' theorem, a theory of related ranges can be built, the
correspondence of two such ranges being without ambiguity when
three points of one are assigned to correspond to three points of
the other. This establishes a corresponding theory for related flat
pencils of lines in a plane, all passing through the same point, and
for related axial pencils of planes in space, all passing through the
same line. And it is, here, further, shewn that there follows from
this a theory of two related planes, in which every point of either
plane corresponds to a point of the other plane, and every line of
either plane, with any range thereon, corresponds to a line of the
other plane, with a range thereon which is related to the former;
this correspondence, of the points of one plane to those of the other,
is determined when *four* points of one plane, of which no three are
in line, are assigned to correspond to four arbitrary points, of
similar generality, in the other plane. Then, further still, it is
shewn that there follows the possibility of a similar unique correspondence of the points of two threefold spaces, in which not only
every point of either space corresponds to a definite point of the
other, and every range of either space corresponds to a related

range of the other space, but also every plane of either space, with the points and ranges thereon, corresponds to a definite plane of the other space which is related to the former plane, as regards its points and ranges, in the manner just described for two related planes; this correspondence is determined when *five* points of one of the two spaces, of which no four are in a plane, and, therefore, no three are in line, are assigned to correspond to five arbitrary points of similar generality in the other space. A similar theory of the correspondence of two n-fold spaces can be deduced, which is determined when $(n+2)$ points of one space, of general position, are assigned to correspond to $(n+2)$ points, of general position, of the other space.

There is, as we shall point out, an exact similarity between the condition necessary to fix the correspondence of the points of two spaces, and the condition necessary to fix the correspondence between the points of either of these spaces and the algebraic symbols by which these points are represented. To fix the correspondence between the points of a line and the symbols by which these points are represented, it is necessary to assign the symbols of three points of the line. This is a consequence of our initial assumption whereby a point by itself is represented as well by a symbol P as by a symbol mP, in which m is an arbitrary algebraic symbol; in what precedes we have frequently employed the remark, having denoted three arbitrary points of a line, respectively, by symbols A, B and $A + B$, thus fixing the multiplier of the symbol B with respect to that of A, and also the multiplier of the symbol of the third point. To fix the correspondence between the points of a plane and the symbols by which these points are represented, it is similarly necessary to assign the symbols of four points of the plane, the points being such that no three of them are in line, and the symbols, therefore, such that no three of them are in syzygy. It is the same for higher cases.

The symbolism is, then, in exact correspondence with the geometry, provided that, (1), to every geometric operation there corresponds a definite operation with the symbols; (2), to every operation with the symbols there corresponds a definite geometrical operation. That the former condition is satisfied follows from the initial descriptions of the use of the symbols; and the latter condition is satisfied for the fundamental laws of operation of the symbols (above, p. 74), each of which has been interpreted geometrically. The Abstract Geometry is not opposed to, but includes, as part of itself, the Real Geometry discussed in the last chapter, the symbols then appropriate being those iterative symbols, and those deduced from them, which correspond to the numbers of ordinary arithmetic. But now the way seems open to us, still further to generalise the Abstract Geometry, with the help of

suggestions arising from the symbols themselves, using the words
point, line, etc., in a proper sense consistent therewith.

We have in fact remarked (p. 66, above) the existence of a system
of algebraic symbols, each formed with two numbers of ordinary
arithmetic, which obey the laws of operation set out above, and are
subject to the commutative law of multiplication; and these contain among themselves symbols corresponding to the numbers of
ordinary arithmetic. These symbols, then, are a system appropriate
to the Abstract Geometry, as so far developed, assuming Pappus'
theorem. This system of symbols has, however, a possibility which
we have not alluded to as yet: *to every symbol c, of the system, there
corresponds a symbol z, such that $z^2 = c$.* In fact, denoting any of
the symbols of this system by (x, y), where x, y are numbers of
ordinary arithmetic, we have $[(0, 1)]^2 = [-1, 0]$, and, more generally,
if $r = (x^2 + y^2)^{\frac{1}{2}}$,

$$[([\tfrac{1}{2}(r + x)]^{\frac{1}{2}}, \ [\tfrac{1}{2}(r - x)]^{\frac{1}{2}})]^2 = (x, y);$$

while, since $(-z)(-z) = z^2 = (z)(z)$, beside any z satisfying $z^2 = c$,
there is also $-z$. Thus, in the geometry to which the symbols of
this system are appropriate, the geometrical operation, previously
given (p. 76), by which we pass from a point $O + zU$ to a point
$O + z^2U$, must be capable of reversion in all cases; we must assign
to the Abstract Geometry such points that, when $O + cU$ is a point,
there is also a point $O + zU$, in which $z^2 = c$. It will be shewn that
a geometrical fact corresponding to this is that, if A, B, C be given
points, and a, b, c be given lines, in a plane containing A, B, C,
then it is possible to draw, in this plane, a line p through the point
A, a line q through the point B, and a line r through the point C,
so that the point (q, r), where the lines q and r intersect, lies on
the line a, and, similarly, the point (r, p) lies on the line b, and the
point (p, q) lies on the line c. Another geometrical fact corresponding to the algebraic condition, which, therefore, is geometrically reducible to the former, requires the notion of related ranges,
which we have developed on the basis of Pappus' theorem. This
theorem has been used to shew that, if two related ranges on the
same line have *three* corresponding points in common, then they
coincide entirely. In the Real Geometry two such related ranges
may, however, have *two* corresponding points in common; for instance, it may be shewn, by the assumptions employed in the last
chapter, that a *sufficient* condition for this, when the points A, B, C
of one range correspond respectively to the points A', B', C' of the
other, is that the order ABC should be opposite to the order $A'B'C'$.
The geometrical fact now referred to, for the case of the Abstract
Geometry represented by the system of symbols now being considered, is that two related ranges on the same line, which do not

coincide entirely, *always have two corresponding points in common*, which, however, may coincide with one another in particular cases. We shall give the geometrical deduction of this geometrical fact from the former, relating to three given points and three given lines. It is clear from what has been proved (cf. pp. 18, 129), that a still further way of stating a geometrical fact equivalent to the above, would be to assume that four lines in threefold space, of which no two intersect, have always two common transversals. Though this is clear by the intervention of the theory of related ranges, it will be proper to reduce it directly to the previously stated theorem for three given points and three given lines. This last theorem, though not considered by Steiner in this general way, we shall refer to, for the sake of brevity, as Steiner's theorem. (See Steiner's *Gesamm. Werke*, Vol. I, pp. 303—305, where reference is made to papers by Servois, Gergonne and L'huilier, in Vol. II, 1811, 1812, of the *Annales de Mathématiques*.)

When we extend the Abstract Geometry in accordance with the suggestions of the system of algebraic symbols above referred to, or in some other way that may prove possible, two questions naturally arise: (1), Is there any geometrical utility in this extension? (2), Is it legitimate to use the postulated properties of the abstract points, lines, etc., in order to prove relations existing among the real points, lines, etc., that is, relations which can be stated without any reference to the abstract elements? The former question is analogous to the question: Is there any utility, for the purposes of the calculus of real numbers, in the introduction of complex numbers? And, as in that case, it may be said, briefly, that experience has amply shewn that the gain in the generality of the statements of geometrical fact, and the increased power of recognising the properties of a geometrical figure, enormously outweigh the initial feeling of artificiality and abstractness[1]. The second question

---

[1] Cayley's remarks may be quoted (Presidential Address, Brit. Assoc., 1883, Collected Papers, xi, 434): "The notion which is really the fundamental one (and I cannot too strongly emphasize the assertion), underlying and pervading the whole of modern analysis and geometry, that of imaginary magnitude in analysis, and of imaginary space (or of space as a *locus in quo* of imaginary points and figures) in geometry."

The opening words of the brief *Vorwort* of von Staudt's *Beiträge zur Geometrie der Lage* (Erlangen, 1856) may also be quoted: "Indem die Mathematik darnach strebt, Ausnahmen von Regeln zu beseitigen und verschiedene Sätze aus einem Gesichtspunkte aufzufassen, wird sie häufig genöthigt, Begriffe zu erweitern oder neue Begriffe aufzustellen, was beinahe immer einen Fortschritt in der Wissenschaft bezeichnet. Dahin gehört namentlich die Einführung von imaginären Grössen in der Analysis und die Einführung von imaginären Elementen in der Geometrie."

Some of the objections raised to the original introduction of complex numbers in Geometry and Analysis are very interesting to read. Many references to the early literature are given by Cayley, *On multiple algebra*, Coll. Papers, xii, 459. A single reference additional to these is Caspar Wessel, *Essai sur la représentation analytique de la direction*, Kopenhagen, 1799.

may perhaps be made clearer by reference to two simple examples: (1) By the use of complex numbers it is easy to prove that a polynomial of the form $x^{2n} - 2x^n \cos n\theta + 1$ is the product of quadratic factors of the form $x^2 - 2x \cos (\theta + 2k\pi/n) + 1$, where n is a positive integer and k has one of the values $0, 1, ..., (n-1)$. In this theorem, x may be a real number, and θ a real angle whose cosine is determined by constructions made with only real points. Is then this result, for real numbers only, legitimately proved by the use of complex numbers? And, if so, does this proof involve that it must be possible to prove the theorem with the use of real numbers only? (as in fact it is). This example, though referring to matters outside our plan, is given as having been of interest to eminent mathematicians. (2) Another example is of directly geometrical importance; but involves some anticipation of succeeding work. As we shall see, for the curve which is called a conic section, if A, B, C, A', B', C' be any six points of the curve, the three point intersections, of pairs of joins of these points, represented, respectively, by $(BC', B'C)$, $(CA', C'A)$ and $(AB', A'B)$, lie on a line. Of this theorem, in fact, Pappus' theorem is a particular case, namely when the conic section degenerates into two lines containing, respectively, the points A, B, C and A', B', C'. It is possible to establish, for the points of a conic section, a theory of related ranges precisely like the theory of related ranges on a line; and, as in that case, three points of a range assigned as corresponding to three given points of another range, suffice to determine the correspondence of other pairs of two points, belonging to these ranges, respectively. Suppose then we establish two related ranges upon the conic, in which the points A, B, C, of one range, correspond, respectively, to the points A', B', C' of the other. Then it may, or may not, happen, that upon the conic, regarded as a locus of real points, there are two points, U, V, each of which, regarded as belonging to one range, corresponds to itself, regarded as belonging to the other range. When these points U, V are existent, as real points, the theorem above stated can be proved at once; the line of the theorem is then, in fact, the join of the points U and V. The question, of which the theorem is to serve as an example, becomes then: Is the theorem then necessarily true, without further proof, when the points U, V do not exist as real points? and can a proof be devised for this case which shall only employ real points? In the Abstract Geometry, the points U, V always exist, and a proof of the theorem may be given identical in form with that which is valid in the real geometry: does this suffice to prove the theorem for the Real Geometry when the points U, V do not exist?

The question is an old one. The reader may compare the discussion of Kepler's ideas, and of the Principle of Continuity, in Ch. Taylor, *Ancient and Modern Geometry of Conics*, Cambridge, 1881, p. lviii, etc.

In a less concrete form, the matter may be stated thus : We have a set of entities, $a, b, c, \ldots$, subject to certain laws of combination; we have also another set of entities, $A, B, C, \ldots, A', B', C', \ldots$, subject to certain laws of combination; these laws will include laws for the combination of $A, B, C, \ldots$ among themselves alone, as well as laws for the combination of the whole aggregate $A, B, C, \ldots$, $A', B', C', \ldots$, when some of $A', B', C', \ldots$ enter. Suppose now that to every law for the combination of $A, B, C, \ldots$ among themselves alone, there corresponds a precisely similar law for the combination of $a, b, c, \ldots$ among themselves, *and conversely*. Then any theorem concerning $A, B, C, \ldots$ alone, which is obtainable logically by the application of the laws of combination which relate to $A, B, C, \ldots$ alone, is equally true for $a, b, c, \ldots$, being obtainable from the laws of combination of $a, b, c, \ldots$ by exactly similar logical processes. Now it may very well happen that a theorem relating to $A, B, C, \ldots$ alone, is easier to prove when regarded as a theorem for particular entities of the whole aggregate $A, B, C, \ldots, A', B', C', \ldots$, the natural method of proof being one which employs not only the laws of combination of $A, B, C, \ldots$ among themselves alone, but also the laws relating to the whole aggregate, using, in its course, one at least of the entities $A', B', C', \ldots$ The question then is, does such a method of proof establish the corresponding theorem for $a, b, c, \ldots$? And can a method of proof be, therefore, devised, employing only the laws of combination of these alone?

In the former of the two examples quoted above, the entities $a, b, c, \ldots$ are the real numbers; the entities $A, B, C, \ldots, A', B', C', \ldots$ are the complex numbers; each of these, we know, is representable as a couple, (x, y), of two real numbers, and among these are the couples $(x, 0)$ whose laws of combination among themselves are precisely similar to those of the real numbers. These couples $(x, 0)$, whose second element is zero, are the entities $A, B, C, \ldots$ of our abstract formulation of the question; the entities $A', B', C', \ldots$ are those complex numbers, (x, y), for which y is not zero, among which is the number $(0, 1)$ whose square is equal to $(-1, 0)$. In the latter of the two examples discussed above, the entities $a, b, c, \ldots$ are the real points, lines, etc., of a Real Geometry, widened by the introduction of the postulated points, lines, etc. The entities $A, B, C, \ldots, A', B', C', \ldots$ are the elements called points, lines, etc. in the Abstract Geometry, among which are elements $A, B, C, \ldots$ (associable with the iterative symbols and those derived from them, which behave like the real numbers of ordinary arithmetic) whose laws of combination and mutual relations (as to order, etc.) are precisely similar to those of the elements of the Real Geometry.

When stated in this abstract form, the question at issue would seem to evaporate. To say that a theorem, which may have been proved with mention of elements $A', B', C', \dots$, relates to $A, B, C, \dots$ alone, must mean that the theorem is a logical consequence of the laws of combination of these among themselves alone; it must therefore be capable of proof with only the use of these laws, provided sufficient ingenuity be forthcoming. That such a constructive faculty of ingenuity is indispensable to any fruitful combination of the preliminary formal laws of combination of the entities of mathematics, is a familiar circumstance of the study of mathematics; and an explicit recognition of the fact should not seem out of place in discussing such a question as that before us. When however the proof is found which can be stated in terms of $A, B, C, \dots$, and their laws of combination, alone, an identical proof is possible for the corresponding theorem relating to $a, b, c, \dots$. That the theorem may have been originally found by considerations in which the elements $A', B', C', \dots$, were involved, as well as $A, B, C, \dots$, thus does not affect the truth of the theorem for $a, b, c, \dots$. But the introduction of the elements $A', B', C', \dots$ may well have assisted the constructive faculty to which reference has been made; that this may happen is, indeed, one of the discoveries of the history of reasoning. The argument is not invalidated by the fact that the elements $A', B', C', \dots$, in their first conception, may be creations of a judicious imagination; so regarded, they may be called abstract elements, and then the elements $A, B, C, \dots$, as forming part of an aggregate containing $A', B', C', \dots$, may equally be regarded as abstract, though framed to have the same relations with one another as have the elements $a, b, c, \dots$.

The importance, and wide bearing, of the matter, seem to call for the indications, which have been given, of the argument by which the attitude we adopt is to be justified. There is however another way in which the use of so-called imaginary elements in geometry may be justified. As complex numbers may be introduced in arithmetic, each as a pair of real numbers, so aggregates of real elements may be introduced in the Real Geometry—if such aggregates can be discovered—whose laws of combination among themselves, and with the existing real elements, shall be those which we ascribe to the imaginary elements. Then such aggregates may be used in place of the imaginary elements; any proposition which we should otherwise state for the imaginary elements, becomes then a proposition for aggregates of real elements, capable of proof by real geometry only. And no doubt arises that a result thereby

10—2

found, relating to the original real elements only, may not be true. It must always remain a striking monument to Karl Georg Christian von Staudt, that he should have elaborated, on a descriptive basis, a theory of aggregates, of real geometrical elements, with the necessary properties (with which should be compared the, frankly metrical and less complete, work of Chasles, *Géométrie supérieure*, esp. Ch. v and Ch. xxxiii). Such a theory does not raise the same logical difficulties as that spoken of above; but, setting aside the necessary prolixity of shewing that the system itself is logical, it should not be out of place to recognise explicitly that the proof of any property of these real aggregates, by the methods of real geometry, is often hard to find; it is often easier to use the language, if not the associated symbolism, of the abstract geometry. When such a theory as von Staudt's has been established this is logically allowable, if it is preferred to such a frank recognition of abstract elements as is suggested above. However this may be, the properties of such aggregates of real elements are interesting in themselves; and we give some account of them, choosing for this purpose a scheme elaborated since von Staudt's time.

Theory of two related plane systems. Suppose two planes, ϖ, ϖ', to be given; and, in one of these, four points, A, B, C, D, which do not lie in line, there being also, in the other plane, four points, A', B', C', D', given, not lying in line. As has been indicated, we agree to say that two flat pencils, each consisting of lines through a point lying in a plane, are related to one another, when the range of points, determined by the former pencil of lines upon an arbitrary transversal, is related to the range similarly determined by the latter pencil upon any transversal. Here, as in what follows, unless the contrary be said, we are assuming Pappus' theorem. If, then, in the former plane, ϖ, any line, l, be drawn through D, a definite line, l', in the plane ϖ', passing through D', is determined, corresponding to l, by the condition that the flat pencil of four lines, $D'A'$, $D'B'$, $D'C'$, l', shall be related to the pencil of four lines, DA, DB, DC, l. In the same way, to any line drawn, in the plane ϖ, through the point A, we can make correspond a definite line, drawn, in the plane ϖ', through the point A'; and similarly for lines drawn through B and B', and for lines drawn through C and C'.

Hence, if P be any point of the plane ϖ, and we draw the lines DP, AP, and then, in the plane ϖ', we draw the lines, through D' and A', corresponding, respectively, to the lines DP, AP, in the sense just explained, we obtain, by the intersection of these, a point, P', of the plane ϖ'. It can be shewn that, then, the lines $B'P'$ and $C'P'$ also correspond, respectively, to BP and CP, in the sense just explained. For, in the plane ϖ, let the lines DA, DP, AP meet the

line *BC*, in the points *X*, *Y*, *Z*, respectively, and let *DP* meet *AB*
in the point *T*. The pencil of lines, with *D* as centre, passing,
respectively, to *B*, *C*, *A*, *P*, is in perspective with the range *B*, *C*, *X*, *Y*.
As this pencil of lines is related to
the corresponding pencil of the
plane ϖ', it follows that the range
B, *C*, *X*, *Y* is related to the range
obtained by similar construction in
the plane ϖ'. The same may be said
of the range *B*, *C*, *X*, *Z*, which lies
on lines from *A* passing, respec-

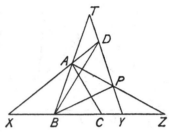

tively, to the points *B*, *C*, *D*, *P*. We have shewn that, if we
assume Pappus' theorem, as we do, here, and in the following
work, two ranges are uniquely related by the assignment of three
points of one to correspond to three points of the other. Thus,
as *B*, *C*, *X*, *Y* and *B*, *C*, *X*, *Z* have the three points *B*, *C*, *X*
in common, it follows that the range *B*, *C*, *X*, *Y*, *Z* is related to
the similarly constructed range of the plane ϖ'. However, the
points of this range, *B*, *Y*, *X*, *Z*, by perspective from *A*, give rise
to the points *T*, *Y*, *D*, *P*, respectively, and these lie on the lines
from *B*, respectively, to the points *A*, *C*, *D*, *P*; the pencil *B* (*A*, *C*, *D*, *P*)
is, thus, related to the similar pencil *B'* (*A'*, *C'*, *D'*, *P'*) of the plane
ϖ'. A similar proof shews that the pencil *C'* (*A'*, *B'*, *D'*, *P'*) is
related to the pencil *C* (*A*, *B*, *D*, *P*).

It can, next, be shewn that to any range of points, lying on a
line of the plane ϖ, there corresponds, by the construction used to
determine *P'* from *P*, a range of points lying on a line of the plane
ϖ', and that this range is related to the former. For this, we shew
that, if *P*, *Q*, *R* be three points of the plane ϖ, lying in line, and if
P', *Q'* be the points of the plane ϖ' which correspond, respectively,
to *P* and *Q*, then there is, upon the line *P'Q'* a point, *R'*, similarly
corresponding to *R*. Let the points where the line *PQ* is met by
the lines *DA*, *DB*, *DC*, respectively, be called *U*, *V* and *W*; and,
similarly, the points where *P'Q'* is met by *D'A'*, *D'B'*, *D'C'* be
called *U'*, *V'* and *W'*. Then, by what has been said, the range
(*U'*, *V'*, *W'*, *P'*) is related to the range (*U*, *V*, *W*, *P*), and the
range (*U'*, *V'*, *W'*, *Q'*) is related to (*U*, *V*, *W*, *Q*); hence the range
(*U'*, *V'*, *W'*, *P'*, *Q'*) is related to (*U*, *V*, *W*, *P*, *Q*). And to the point
R, of the line *PQ*, regarded as a point of the range (*U*, *V*, *W*, *P*, *Q*, *R*),
will correspond a point, R_1', of the line *P'Q'*, such that the range
(*U'*, *V'*, *W'*, *P'*, *Q'*, R_1') is related to the former; we are to shew
that this point R_1' is the same as the point, *R'*, constructed, with
the help of *A'*, *B'*, *C'*, *D'*, to correspond to *R*. Let *AB*, *AC* meet
the line *PQ*, respectively, in *Y* and *Z*, and *A'B'*, *A'C'* meet *P'Q'* in

Y', Z'. Then, by the original construction, by perspective from A, the range (U', P', Q', Y', Z') is related to (U, P, Q, Y, Z); and,

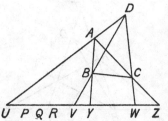

therefore, as three points suffice to identify a range which is related to another, the range $(U', V', W', P', Q', R_1', Y', Z')$ is related to (U, V, W, P, Q, R, Y, Z). Thus the point R_1' satisfies the condition which is to be satisfied by R' with regard to pencils of lines whose centre is A', and also with regard to pencils of lines whose centre is D'. These two are sufficient to identify R_1' with R'. There is thus a range $P'Q'R'...$ in the plane ϖ' which is related to the arbitrary range $PQR...$ in the plane ϖ, as we desired to prove. From this it is easy to deduce that there is a pencil of lines, passing through a point, in the plane ϖ', which is related to any such pencil of the plane ϖ.

The plane systems, in ϖ and ϖ', so constructed, will be said to be *related*. The words *homographic*, and *projective*, are also used later on; but as they are elsewhere used with implications upon which we may not have entered as yet, their use is postponed.

Two related threefold systems, or spaces. A similar correspondence may be set up between the points, lines and planes of two three-dimensioned spaces, Σ and Σ', by taking any five points A', B', C', D', E' of the space Σ', of which no four lie in a plane, to correspond, respectively, to five arbitrary points A, B, C, D, E, of similar generality, in the space Σ.

There is a preliminary point to which a word may be given. In establishing a correspondence between two planes, we have pre-supposed a construction possible by which ranges on two lines, in these planes respectively, may be related to one another; this is certainly so when the planes lie in the same threefold space. But in comparing two threefold spaces, the presupposition must be a definite one; in the absence of a briefer general method, we may suppose the two threefold spaces to be both contained in a higher space, and ranges lying therein to be related by a chain of per-spectivities, as in the foregoing theory, in Chapter I. When ranges belonging to the two spaces have been related, plane pencils of lines, and axial pencils of planes, can be related, by the ranges they determine on a transversal. In particular, however, the two three-fold spaces may be the same.

This being understood, let ϖ be any plane through the line BC of the space Σ; with the planes joining BC to the points A, D, E, we then have an axial pencil of four planes. The planes, in the space Σ', which join $B'C'$ to A', D', E' respectively, taken with a plane, ϖ', through $B'C'$, in this space, form another axial pencil; and the plane ϖ' is determined to correspond to ϖ when we suppose it taken so that these two axial pencils of planes are related. Now let the line DE meet the plane ABC in the point D_1, and, similarly, the line $D'E'$ meet the plane $A'B'C'$ in the point D_1'; then, suppose the planes ABC and $A'B'C'$ related to one another, as in the preceding article, by the condition that the points A, B, C, D_1 shall correspond, respectively, to A', B', C', D_1'; and let P' be the point of the plane $A'B'C'$ so corresponding to a point P of the plane ABC. Next, let the planes BCE and $B'C'E'$, in the spaces Σ and Σ', meet DP and $D'P'$, respectively, in Q and Q'; and, corresponding to any point R taken upon the line DP, let R' be the point of the line $D'P'$ determined by the condition that the ranges $P'Q'R'D'$, $PQRD$ shall be related. Then R' is such that the two axial pencils, each of four planes, in the spaces Σ' and Σ, represented, respectively, by $B'C'$ (A', E', R', D') and BC (A, E, R, D), with axes $B'C'$ and BC, are related to one another. The axial pencils represented, respectively, by $A'D'$ (B', C', E', R') and AD (B, C, E, R) are also related, since, on the planes $A'B'C'$ and ABC, these give, respectively, the flat pencils A (B, C, D_1, P) and A' (B', C', D_1', P'), of centres A and A'; and, by construction, the plane systems, (A, B, C, D_1, P) and (A', B', C', D_1', P'), are related. The five given points A, B, C, D, E lead to ten lines joining pairs of these; by passing through each of these joining lines a plane, to the other three of the five given points, and to the point R, we obtain eight axial pencils, each of four planes, beside the two we have considered having BC and AD for axes. It can be shewn that every one of these is related to the axial pencil similarly constructed in the space Σ'. Consider, for instance, the axial pencil CA (B, D, E, R), of axis CA. The point P is on the plane ABC, and the point Q is on the plane BCE; let the line DP meet the plane ACE in the point X. The axial pencil CA (B, D, E, R) gives, on the line DP, the range (P, D, X, R). The line ED meets the plane CAB in the point D_1. Thus, the range (P, D, X, Q), being a section of the axial pencil CE (P, D, A, B), of which a plane section is the flat pencil C (P, D_1, A, B), is related to this flat pencil; and this, by construction, is related to the corre-

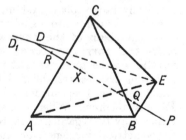

sponding flat pencil C' (P', D_1', A', B') of the space Σ'. Wherefore the range (P, D, X, Q) is related to the corresponding range $(P', D'X', Q')$ of the space Σ'. By definition of R', however, the range (P, D, Q, R) is related to the corresponding range (P', D', Q', R'); as (P, D, X, Q) and (P, D, Q, R) have the three points P, D, Q in common, we infer that the range (P, D, X, Q, R) is related to the corresponding range (P', D', X', Q', R'). We have seen, however, that the range (P, D, X, R) is a section of the axial pencil $CA\ (B, D, E, R)$. This axial pencil is thus related to the corresponding axial pencil $C'A'\ (B', D', E', R')$, of the space Σ', as we desired to shew. The range (D, X, Q, R) is also a section of the axial pencil $EC\ (D, A, B, R)$, which is, therefore, related to the corresponding axial pencil $E'C'\ (D', A', B', R')$.

The point R is any point of the line DP, and P is any point of the plane ABC; thus the point R is any point of the threefold space A, B, C, D, E. With the proof of the theorem above stated, that the ten axial pencils of planes whose axes are the joins of pairs of the points A, B, C, D, E, each pencil containing a plane passing through R, are related, respectively, to the corresponding axial pencils in the space Σ', there is thus established a correspondence between the points of the spaces Σ and Σ'. It can then be shewn, in a manner analogous to that adopted for the case of two planes, that a range of four points lying in line in the space Σ corresponds to a related range of four points lying in line in the space Σ'; and then that a set of five points lying in a plane in the space Σ gives rise to a set of five points lying in a plane in the space Σ', in such a way that the pencil of lines joining one of the five points to the other four, in Σ, is related to the corresponding pencil in Σ'.

Case of two spaces both of n dimensions. And the argument can, it appears, be extended, by the method of induction, to two spaces of any number of dimensions, it being necessary and sufficient, in order that two n-fold spaces should be related, that $(n + 2)$ points of one space, of general position, be assigned to correspond to $(n + 2)$ points of the other, also of general position. Then any range, plane system, threefold system, ..., contained in one space, is related to a similar system of the other space.

Comparison of the correspondence between two spaces with the correspondence between either space and the symbols used to represent it. The correspondence in the three foregoing articles is established by purely geometrical considerations, based on the Propositions of Incidence, and the assumption of Pappus' theorem. When two spaces are so related to one another, either may be regarded as representative of the other, there being a geometrical incidence, or theorem, in either, corresponding to any geometrical incidence, or theorem, in the other. Conversely two spaces, of a like number of dimensions, with a point to point correspondence of such a kind that, to every incidence of elements in one space, corresponds a precisely similar incidence of elements in the other, are necessarily related, as we see by recalling the construction of two related ranges. Now compare, with this correspondence between two spaces, the correspondence we have sought to set up between one of these spaces, and the realm of the symbolism we have introduced to represent this space (Chap. I, Section III). To every element, or incidence of elements, in the space, corresponds a symbol, or agreement of two symbols, in the algebraic realm; and, conversely, to every fundamental identity (law of operation) between the symbols, and, hence, to every derived identity, corresponds a coincidence of points in the space, as we have shewn in detail. Moreover the *freedom* in the establishment of the correspondence between the related spaces, that $(n + 2)$ points of one space may be chosen arbitrarily to correspond to $(n + 2)$ points of the other, agrees with the circumstance that, to fix the symbol of every point of the space, the symbols of $(n + 2)$ of its points must be assigned, although these $(n + 2)$ symbols will be in syzygy. This condition is, in part, to be compared with the fundamental proviso that the dimensions of the two spaces must be the same; but it secures further that the points referred to, in the statement that every identity of symbols corresponds to a coincidence of points, shall be without ambiguity; without this condition, there would be the same want of definite correspondence between the elements in the space and in the algebraic realm as exists between the elements of a space and every other which is related thereto.

For the points of a line, in order to be able to assign a definite symbol to every point, it is necessary that the symbols belonging to three points of the line should be assigned. If A, B be two points of the line, the former of these may, by the rules given for the symbols, be represented either by a symbol A, or by a symbol mA, where m is any one of the algebraic symbols. Thus the symbol for a general point of the line, $A + xB$, is incapable of identifying a particular point, since it might equally have arisen in the form $mA + xnB$, where m, n are any two of the algebraic symbols; this is the same as $A + m^{-1}nxB$, since, with Pappus' theorem, the algebraic symbols are commutative in multiplication. If, however, the symbol be given which is to belong to a third definite point of the line, say $A + x_0B$, with a definite symbol for x_0, this suffices to fix the previously unspecified factor $m^{-1}n$. Then every point of the line has a definite symbol; and, conversely, we have shewn how to construct on the line the point corresponding to every symbol $A + xB$ which can arise by the operations to which these algebraic symbols are liable. That the condition secures the same kind of definiteness as is involved in distinguishing between two related ranges, may be well illustrated by proving the following result, of which frequent use is made subsequently : *The condition that four points of a line represented, respectively, by symbols A, B, A + xB, A + yB, should be a range related to that of four points, on the same or on another line, which have symbols P, Q, P + pQ, P + qQ, is that* $yx^{-1} = qp^{-1}$. In accordance with our conventions, this is the same, writing B for xB, and Q for pQ, as saying that the condition that the range of four points represented, respectively, by $A, B, A + B,$ $A + \xi B$, should be related to the range of four points represented, respectively, by $P, Q, P + Q, P + \sigma Q$, is that $\xi = \sigma$. This is in accordance with the comparison we have made between the condition that two ranges should be related, and the condition that the points of a range should be represented by the symbols. To prove the result, when the two ranges are on non-intersecting lines in three dimensions, let the four joins of the corresponding points, of A to P, of B to Q, of $A + B$ to $P + Q$, and of $A + \xi B$ to $P + \sigma Q$, be met by another line, respectively, in points with symbols $A + \lambda P$, $B + \mu Q$, U and V. As the point U is then a derivative of $A + B$ and $P + Q$, as well as of $A + \lambda P$ and $B + \mu Q$, we infer $\lambda = \mu$, and $U = A + B + \lambda (P + Q)$. As V is a derivative as well of $A + \xi B$ and $P + \sigma B$ as of $A + \lambda P$ and $B + \lambda Q$, the symbol $A + \xi B + \kappa (P + \sigma B)$ must, for proper κ, be the same as $A + \lambda P + \xi (B + \lambda Q)$, so that $\kappa = \lambda$, and hence $\lambda \sigma = \xi \lambda$. This gives[1], as stated, $\sigma = \lambda$. The symbols are assumed commutative in multiplication, without which, as was shewn in Chapter I, there is not definiteness in the condition that two ranges be related.

For the points of a plane, to render unique the correspondence between the symbols and the points, it is necessary to distinguish between the symbols A, B, C of three fundamental points of the plane, and, respectively, the symbols of the forms lA, mB, nC, which represent the same points. And this can be done by providing that in the symbol $xA + yB + zC$, belonging to a chosen definite point of the plane, the symbols $x^{-1}y, x^{-1}z$ shall have stated values. Very often the most convenient way of doing this is, to select the point to which the symbol $A + B + C$ is to be attached. As will be seen later, in dealing with the theory of linear transformations, the plane system of which the points are represented by symbols $xlA + ymB + znC$, wherein l, m, n are the same for all points, is related to the plane system of which the points are represented by symbols $xA + yB + zC$, wherein x, y, z vary from point to point.

---

[1] So that, as was remarked (above, p. 93), when approached from this point of view, von Staudt's introduction of numbers, to represent Würfen, or ranges of four points on a line, involves the assumption of Pappus' Theorem. For, without this Theorem, we have shewn that the range $B, A, B + \mu^{-1} A, B + \lambda^{-1} A$ is related to the range $A, B, A + \lambda B, A + \mu B$; and $\mu \lambda^{-1} = \lambda^{-1} \mu$ requires $\lambda \mu = \mu \lambda$.

Similar remarks evidently apply whatever be the number of dimensions of the space considered.

A geometrical existence theorem for the solution of an algebraic equation. The preceding article is concerned with the justification of the use of algebraic symbols of any generality which are subject to the laws of combination set out in detail in Section III of Chapter I, together with that of commutative multiplication. As indicated in the discussion at the beginning of this chapter, there is a quite definite system of symbols with these properties which has the further property of containing, corresponding to any one of its members, c, another member, z, which is such that $z^2 = c$. It is desirable then to examine the geometrical meaning of this. If this system of symbols be that which is adopted as guide in formulating the properties of the *points*, *lines*, etc., which we finally adopt as the elements of the Abstract Geometry, it is convenient to summarise these properties, if possible, by enunciating a geometrical construction which can be carried out with the abstract elements, or by stating a theorem which holds when these elements are allowed.

With this object we turn now to a geometrical problem, which we consider, in the first instance, in an elementary way; shewing how it is related to other problems whose fundamental importance may be, at first sight, more obvious:

In a plane, we are given three points, A, B, C, and three lines, a, b, c. It is required to draw three lines, p, q, r, passing, respectively, through the points A, B and C, whose points of intersection in pairs, namely P, of q and r, Q, of r and p, and R of p and q, shall lie, respectively, on the lines a, b, and c. In this case, the lines joining the pairs of points, namely p, of Q to R, q, of R to P, and r, of P to Q, pass, respectively, through the points A, B, and C. Thus the problem is one of a self-dual character.

We arrange the considerations, which are given, under nine headings:

(i) There is an infinite number of solutions of the problem when the lines BC, CA, and AB, pass, respectively, through the points (b, c), (c, a), and (a, b). This is, in fact, another statement of Pappus' theorem.

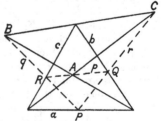

It can be shewn, when no one of the points A, B, C lies on any of the lines a, b, c, that this is the only case in which there is an infinite number of solutions of the problem.

(ii) When the points A, B, C are in line, a degenerate solution of the problem is that in which the

lines p, q, r all coincide in this line; in this case, another solution
can be directly constructed. Reciprocally, when the lines a, b, c
meet in a point, a degenerate solution is that in which the lines
p, q, r all pass through this point; in this case, another solution
can be directly constructed.

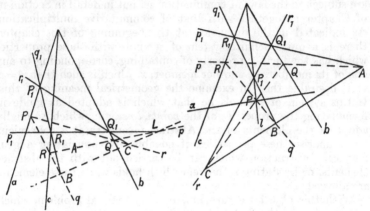

In the first case, when the points A, B, C are in line, draw
through A any line, p_1, meeting the lines b and c, respectively, in
Q_1 and R_1, and let the lines CQ_1 and BR_1 intersect in the point P_1.
Make a similar construction with a line, p_2, drawn through A, so
obtaining a point P_2. Then, let the line P_1P_2, or l, which contains
(b, c), meet the line a in the point P. It will be seen that, if the
lines CP, BP be now drawn to meet the lines b and c, respectively,
in Q and R, then the line QR passes through the point A, and we
have a solution of the problem. A similar construction is valid for
the reciprocal case, starting with points, P_1 and P_2, taken anywhere on the line a; the lines, p_1 and p_2, reciprocal to P_1 and P_2,
intersect in a point L, lying on BC; the required line, p, is the
line AL.

The proof of the statements made is immediate from Desargues'
theorem. In the former case, the corresponding joins of points of
the two triads P_1, Q_1, R_1 and P_2, Q_2, R_2, meet in the points A, B, C,
respectively, which are in line. Hence the join P_1P_2 passes through
the point (b, c), which is the intersection of Q_1Q_2 and R_1R_2. And
then, as the lines, PP_1, QQ_1 and RR_1, meet in the point (b, c), it
follows, from the triads P, Q, R and P_1, Q_1, R_1, that the lines QR
and Q_1R_1 meet on the line BC, in the point A. The proof in the
latter case is the reciprocal of this.

A particular case of the figure arises when the points A, B and C
lie, respectively, on the lines a, b and c. Then it is easy to see that

the points P, Q, R are harmonic conjugates, respectively, of A, B, C, in regard to the respective couples of lines b and c, c and a, a and b. If the line ABC be denoted by p', then the lines p and p' are harmonic conjugates in regard to the line a and the line joining the points A and (b, c), etc.

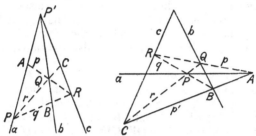

(iii) The simplest, and, as will be seen, the most fundamental of the cases in which the lines p, q, r cannot be directly constructed, is that, in which the points A, B, C lie, respectively, on the lines a, b, c, but are not in line. When this is so, it can be seen, by considerations of betweenness, such as are used in Chapter II, that, in the Real Geometry, there exist lines p, q, r provided the points A, B, C have suitably limited positions. If the points (b, c), (c, a) and (a, b) be denoted, respectively, by U, V and W, this is so when A, B and C are outside the respective segments VW, WU and UV; or when one of them is outside and the other two are inside these segments, respectively.

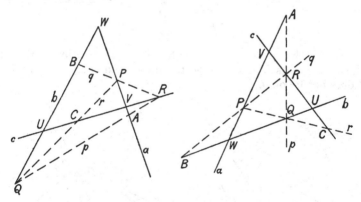

For in the former case, if two lines, p_1 and p_2, be drawn, through the point A, to meet the line b, respectively, in the points Q_1 and Q_2, and these meet the line c, respectively, in R_1 and R_2, in such a way that UR_1R_2V are in order, then also UQ_1Q_2W will be in order. If then BR_1 and CQ_1 meet the line a, respectively, in H_1 and K_1, and BR_2, CQ_2 meet a, respectively, in H_2, K_2, it follows that WH_1H_2V are in order, and that WK_2K_1V are in order. By use of the Dedekind suggestion, as in Chapter II, it can thence be shewn that there is a position of R_2 for which H_2 and K_2 coincide, say in P_2. A similar argument is applicable to the case, for instance, in which B is within the segment WU, and C within the segment UV, but A is without the segment VW, and to the other two such cases. The reciprocal figure, it is easily seen, leads, essentially, to the same results.

When we have one possible set of positions for P, Q, R, solving the problem, another set can be deduced at once; namely, by taking, in place of P, the point, P', which is the harmonic conjugate of P in regard to V and W, or, say, $P' = (V, W)/P$, and taking, at the same time, $Q' = (W, U)/Q$ and $R' = (U, V)/R$.

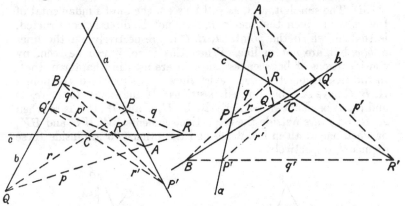

(iv) Another case which is worthy of special mention is that in which the points B and C lie on the lines b and c, respectively, but A does not lie on a.

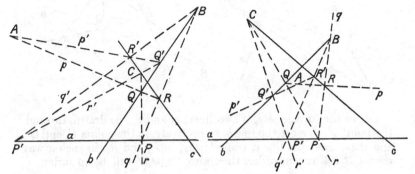

It will be seen below that the solution in this case can be constructed from that considered under (iii).

(v) The case in which the points A, B, C have arbitrary positions can be reduced to that preceding, in which two of the points, B and C, lie on their respectively associated lines, b and c.

The given lines a, b, c, and the given points, B, C, define a line, m, as follows: taking arbitrary points H, H', H'', ..., on the line a, let the joins of these to the point B, namely the lines HB, $H'B$, $H''B$, ..., meet the line b in the respective points K, K', K'', ..., and the joins of these to the point C, namely the lines HC, $H'C$, $H''C$, ..., meet the line c in the respective points L, L', L'', ...; the two ranges, K, K', K'', ..., 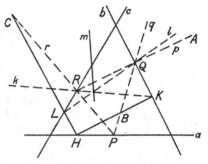 and L, L', L'', ..., are then related. Therefore, as was proved in Chapter I, the points of intersection of the various pairs of cross joins of these ranges, namely, the points $(KL', K'L)$, $(KL'', K''L)$, $(K'L'', K''L')$, ..., all lie on a line which we denote by m.

Take two corresponding points, K and L, of these two related ranges, lying, respectively, on the lines b and c, and, with these, the point A. Take the lines b, c and m. By the solution of the previous case, (iv), we can draw through K, L and A, respectively, the lines k, l and p, so that their intersections in pairs shall lie, respectively, on the lines b, c and m, namely the point (l, p) on b, the point (p, k) on c and the point (k, l) on m. Hence, if the points (l, p) and (p, k) be, respectively, Q and R, these will be points of the ranges $(K, K', ...)$ and $(L, L', ...)$ above described, and the lines BQ, CR, which we may call, respectively, q and r, will meet in a point, P, of the line a. The lines p, q, r are then such lines as are required for the solution of the general case of the problem in hand, which is then reduced to the case when B and C lie on b and c respectively, here considered under (iv).

(vi) There is another problem of fundamental importance which is also reducible to the case considered under (iv), that, namely, of the construction of the common corresponding points of two related ranges lying on the same line.

Let $(B, C, ...)$ and $(B', C', ...)$ be two ranges on a line, p, of which the latter is related to the former, according to the definition we have adopted in Chapter I (above, p. 25), by the fact that $(B', C', ...)$ is in perspective, from a point O, with a range, $(Q, R, ...)$, of another line, q, which is related to the range $(B, C, ...)$. The line

q will intersect p. Thus, as was proved for two related ranges on different lines in the same plane, the cross joins BR, CQ meet in a point lying on a definite line, r, whatever pair of corresponding points of these ranges is substituted for C and R.

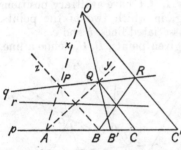

A point, A, of the range $(B, C, \ldots)$, coincides with the corresponding point, A', of the range $(B', C', \ldots)$, if, and only if, the corresponding point, P, of the related range on q, lie with A on a straight line through O. Thus, the problem of finding A is that of drawing through the point O, and through two, arbitrarily taken but corresponding, points Q and B, lying, respectively, on the lines q and p, the respective lines x, y and z, so that y and z may meet on r, z and x may meet on q, and x and y may meet on p.

This reduces the problem of finding the common corresponding points to that considered under (iv). For this problem, reference may also be made to a solution of entirely different character, Chasles, *Géom. supér.* (1880), p. 175.

(vii) The problem of finding common transversals of four lines in three dimensions, of which no two intersect, is also reducible to that considered under (iv).

Let a, b, c, d be four lines, in three dimensions, of which no two intersect. Take any two arbitrary fixed points, O and U, upon the line a. Draw from O the transversal to the lines b and d, meeting d in B; also from O the transversal to c and d, meeting d in Q. Draw from U the transversal to the lines b and d, meeting these, respectively, in V and C; also from U the transversal to c and d, meeting these, respectively, in R and C'. The plane VUR contains the line d; the plane from O containing the line b meets the plane VUR in the line VB, and the plane from O containing the line c meets the plane VUR in the line RQ. Now take any point, H, of the line a: from H draw the transversal to the lines b and d, meeting these, respectively, in M and A, and also the transversal to the lines c and d, meeting these, respectively, in N and A'. The plane from M containing the line a will meet the plane VUR in the line UA; this line will, therefore, intersect the line VB, say in the point T, and the line TM will intersect the line a; the line TM lies, however, in the plane MVB, which meets the line a in the point O; thus, the points O, M, T are in line. The plane from N containing the line a will meet the plane VUR in the line UA'; this line will, therefore, intersect the line RQ, say in the point P, and the line

PN will intersect the line *a*; the line *PN* lies, however, in the plane *NRQ*, which meets the line *a* in the point *O*; thus, the points *O, N, P* are in line.

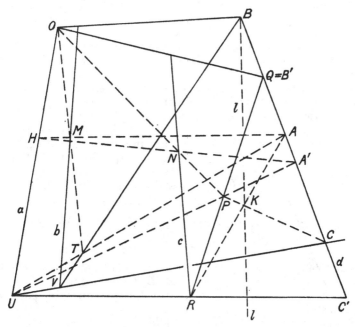

As the point *H* takes different positions on the line *a*, the points *A′, P, N, M, T, A* will take different positions, each on a line independent of *H*, constituting ranges which are all related.¹ In particular, when *H* is at *O*, the points *A′, P, A* will be, respectively, at *B′, Q, B*, on the line *d*, the points *B′* and *Q* being the same; and, when *H* is at *U*, the same points *A′, P, A* will be, respectively, at *C′, R, C*. Considering the two related ranges, in one plane, whose points are the corresponding positions of *P* and *A*, respectively, of which two corresponding points are, respectively, *R* and *C*, arising when *H* is at *U*, the cross joins, *PC* and *AR*, meet in a point, *K*, which describes a definite line, *l*; and this line contains the intersection of the cross joins, *RB* and *CQ*, which is the point *B*; for *Q* and *B* are simultaneous positions of *P* and *A*, arising when *H* is at *O*.

In order, now, to draw a common transversal of *a, b, c, d*, it is necessary to find such a position for *H* that *A* and *A′* may coincide. For this it is necessary, and sufficient, to draw from the given points *U, R* and *C*, of the plane *VUR*, three lines *UPA′, RA, CP*, respectively, lying in this plane, so that the intersections of pairs of these, (RA, CP), (CP, UPA'), (UPA', RA), shall lie, respectively, on the given lines l, RQ, $C'Q$, of this plane. Of the given points and lines, the points R and C lie, respectively, on RQ and $C'Q$. The problem of drawing the common transversals of the four given lines a, b, c, d, is, thus, reduced to the problem above considered, under (iv). For this problem, the reader may also consult Steiner, *Ges. Werke*, I, p. 403, where reference is made to Gergonne's *Ann. de Math.* XVII (1826, 7), p. 83.

(viii) Consider now the problem above referred to as (iii), in connexion with the symbolism which we have used.

Take the points (b, c), (c, a) and (a, b) as fundamental, the lines

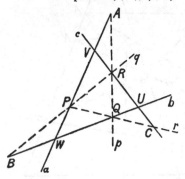

a, b, c being supposed not to meet in a point. Denoting these points, respectively, by U, V, and W, we may choose the symbols V and W, in relation to U, so that

$$C = U - V, \quad B = U - W,$$

and then take, for the symbol of the point A,

$$A = a^2V - W,$$

where a^2 is not 1, the points A, B, C not being in line. If then the symbol of the point P be taken as

$$P = \lambda V + W,$$

this involves in turn, for the symbols of the points Q, R and A, respectively

$$Q = \lambda U + W, \quad R = U + \lambda V, \quad A = \lambda^2 V - W.$$

Identifying this form for A with that given, we infer $\lambda^2 - a^2 = 0$, or $(\lambda - a)(\lambda + a) = 0$, and it was definitely assumed that the system of symbols used should have only 0 as singular symbol. Wherefore λ is a or $-a$. Thus, when the given points A, B, C are, respectively, written as $a^2V - W$, $U - W$ and $U - V$, one set of positions, for P, Q, R, is that given by

$$P = aV + W, \quad Q = aU + W, \quad R = U + aV,$$

another, and the only other, set being obtained from this by putting $-a$ in place of a.

It is seen without difficulty that the figure employed here agrees with the construction given in Chapter I (above, p. 84) for determining the point $O + a^2U$ when $O, U, O + U$ and $O + aU$ are given.

(ix) Take now the case, considered in (iv), in which the given points B and C lie respectively on the lines b and c.

As before, let the points (b, c), (c, a) and (a, b) be, respectively, denoted by U, V and W, and suppose neither B nor C to be at U. We can, then, suppose the symbols of A, B and C to be, respectively, of the forms

$$B = U - 2W, \quad C = U - 2V,$$
$$A = U + mV + nW.$$

If then the symbol of the point P be taken as

$$P = mV + \lambda W,$$

those of Q and R will, respectively, be

$$Q = mU + 2\lambda W, \quad R = U + 2\lambda^{-1} mV,$$

and, in order that the line QR should pass through the point A, there must exist two symbols, q and r, securing the syzygy

$$q(mU + 2\lambda W) + r(U + 2m\lambda^{-1} V) = U + mV + nW.$$

This requires the equations

$$qm + r = 1, \quad 2rm\lambda^{-1} = m, \quad 2q\lambda = n;$$

we are assuming Pappus' theorem, so that the symbols are commutative in multiplication; hence, eliminating q and r, we deduce

$$mn\lambda^{-1} + \lambda = 2, \quad \lambda^2 - 2\lambda + mn = 0.$$

Thence, if θ be a symbol such that

$$\theta^2 = 1 - mn,$$

we have $(\lambda - 1)^2 = \theta^2$, and hence λ has one of the two forms

$$\lambda = 1 + \theta, \quad \lambda = 1 - \theta.$$

Now, by the constructions given in Chapter I (pp. 74, 76), when the points A, B, C, and therefore the symbols m and n, are given, we can construct the point whose symbol is

$$mV + (1 - mn) W,$$

or $mV + \theta^2 W$. From this, by the construction under (viii), preceding, we can find the two points with symbol

$$mV + \theta W,$$

and hence the points of symbol

$$mV + (1 + \theta) W.$$

The general problem now under consideration can, therefore, be solved by direct construction if that under (iii) or (viii) can always be solved.

Remark. The reader may like to see these constructions carried out, in part. The given points are U, V, W, which are the points (b, c), (c, a), (a, b), together with $A = U + mV + nW$, $B = U - 2W$, $C = U - 2V$. A construction for

$mV+(1-mn)\,W$ is as follows: By joining $U+mV+nW$ to V and W we obtain, respectively, $U+nW$ and $U+mV$, on the lines UW and UV. By joining these, respectively, to $U-2V$ and $U-2W$, we obtain $2V+nW$ and $mV+2W$, on VW. From the former of these we construct $V+nW$, the harmonic conjugate of V in regard to $2V+nW$ and W; from the latter we construct $mV+W$, the harmonic conjugate of W in regard to $mV+2W$ and V. From $V+nW$, by taking its harmonic conjugate in regard to V and W, we

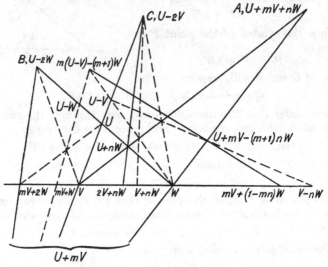

obtain $V-nW$. Then, having $V+m^{-1}W$ and $V-nW$, we find $V+(m^{-1}-n)\,W$, or $mV+(1-mn)\,W$, by the rule for constructing a point specified by the sum of two algebraic symbols which was given in Chapter I. Namely, two lines through $mV+W$ and $V-nW$, respectively, are met by a line through V in the points $U+mV$ and $U-V$, and lines through W to these points meet the two lines first drawn, alternatively, in $U+mV-(m+1)\,nW$ and $m(U-V)-(m+1)\,W$; the join of these gives the point $mV+(1-mn)\,W$.

This is the point $mV+\theta^2 W$, on the line VW; and we have $U-W$ on the line UW, the harmonic conjugate of W in regard to U and $U-2W$, and $mV+U$ on the line UV. Hence, if the result of (viii), preceding, be assumed, we can find $mV+\theta W$, on the line VW. Then, having $mV+W$, we can construct $mV+(1+\theta)\,W$.

Statement of the further Axiom henceforth assumed in the Abstract Geometry. In accordance with the view here adopted, of the possibility of extending the conception of the geometrical entities discussed, we shall suppose, in future, that, in the Abstract Geometry, the meanings attached to the words *point, line, plane*, etc., are such that the system of algebraic symbols appropriate to their discussion, beside obeying the formal laws of combination set forth in Chapter I, and the commutative law in multiplication, is also such that, to any symbol, c, there exists another, z, such that $z^2 = c$. From the preceding discussion it follows that this is equivalent to supposing that the geometrical entities, beside being subject to the Propositions of Incidence, and such that Pappus' theorem is true, are also such that a certain geometrical problem is always capable, if not of an infinite number, then, of *two*, solutions, which may coincide. We have given three alternative forms for this problem, either of which may be taken, in virtue of the other assumed properties of the entities. Either we may say that, if A, B, C be points given anywhere on three lines a, b, c, lying in a plane, it is possible to draw three lines p, q, r, respectively through A, B, C, such that the points of intersection (q, r), (r, p), (p, q) lie, respectively, on a, b, c. From this it follows that the same is true for quite arbitrary positions of A, B, C, not lying on the lines a, b, c. Or we may say that two related ranges of points on the same line, not entirely identical, have two common corresponding points, which may coincide. Or, finally, we may say that, four lines in space of three dimensions, of which no two intersect, are met, if not by an infinite number, then, by *two* transversals, which may however coincide.

It may be said that, later on, it will be assumed that the geometrical entities are such that, in the symbolism appropriate thereto, equations of general character, not merely $z^2 = c$, are capable of solution. It would be proper to discuss now whether this is an additional condition, or contained in what has been adopted. But we do not enter into this at present.

Clearly, from the preceding discussion, the problem of which we postulate a solution in the Abstract Geometry, has not, in all cases, a solution in the Real Geometry. When the theory is supposed to deal only with a real geometry, it is, therefore, usual to speak of *imaginary* elements. As we have said, we may, however, introduce into the Real Geometry certain aggregates of real elements, whose relations with the primary real elements, and with one another, are formally the same as those of the so-called imaginary elements. We now pass to a brief account of aggregates of this kind which may be taken, if it be desired to maintain the point of view of a Real Geometry.

Aggregates of real elements with the properties of imaginary elements. We desire to avoid the use of the words *imaginary point*, *imaginary line*, etc., when establishing the theory, at least. We shall, therefore, use a phraseology which is, confessedly, somewhat cumbrous. In addition to points, we consider *Point-Sets*, each of three points on a line : Any three points on a line, A, B, C, are regarded as taken in one of two definite orders, the sets B, C, A and C, A, B being regarded as equivalent with the set A, B, C; and the sets C, B, A and A, C, B as equivalent with the set B, A, C. Then, further, the set A, B, C is regarded as equivalent with any other set, A', B', C', of three points of the same line, when the three ranges (A', A, B, C), $(B' B, C, A)$, (C', C, A, B) are related. Thus, Pappus' theorem being assumed, a set, A', B', C', can be found equivalent with the set A, B, C, of which one point, say A', is any arbitrary point of the line, the points B', C', associated therewith, being then determinate. It is easy to prove that the ranges (A, A', B', C'), (B, B', C', A'), (C, C', A', B') are related, so that if the set A', B', C' is equivalent to the set A, B, C, then this latter is equivalent to the former. It can be further shewn that two sets, A', B', C' and A'', B'', C'', which are both equivalent to A, B, C, are equivalent to one another.

The whole aggregate of equivalent point sets of the line is the entity called a *point series*. From three points of the line, taken in different orders, two such point series can be formed; these may be called *conjugate* point series. When two of the points of a set A, B, C, of a point series, are coincident, say A with B, the point sets, equivalent with this, are those formed with A, B and any further arbitrary point of the line, and one particular set, of these, consists of the point A taken three times over. In the Real Geometry we have, upon the line, points; in the scheme we are explaining we have, *general point series*, each determined by a set of three points of which no two coincide, and *the degenerate point series*, each of which is determined by three coincident points; while a general point series lies on a definite line, an infinite number of lines contain the three coincident points of a degenerate point series. The laws of manipulation of the general point series will lead to laws for the degenerate point series which are exactly similar to those of ordinary real points[1]. We may say that the number of point series upon a line is ∞^2; for of a point set determining the series, one point may be at any position chosen beforehand, the same for all series, and any particular series then depends on the positions of the other two points of a set determining the series; each of these two points is capable of positions on the line which we usually say to be ∞^1 in number.

Reciprocal to point sets, each of three points on a line, we consider also plane sets, each of three planes passing through a line. A set of three such planes, α, β, γ, is regarded as equivalent to either of the two sets β, γ, α or γ, α, β, but distinct from the equivalent sets, γ, β, α, or β, α, γ, or α, γ, β. And a set of three other planes, α', β', γ', passing through the same line, is regarded as equivalent to α, β, γ when the axial pencils $(\alpha', \alpha, \beta, \gamma)$, $(\beta', \beta, \gamma, \alpha)$, $(\gamma', \gamma, \alpha, \beta)$ are related. The aggregate of all equivalent plane sets is called a *plane series*; and there are degenerate plane series in

---

[1] The reader may compare the general logical formulation given above, p. 146.

which all the equivalent plane sets have two coincident planes in common.

A point series is then said to *lie upon* a plane series, when the points of any one of the equivalent sets of the point series, taken with regard to order, lie, respectively, upon the corresponding planes of any one of the equivalent sets of the plane series. In particular the condition is satisfied if the points A, B, C lie, respectively, upon the planes β, γ, α, as well as if they lie respectively upon α, β, γ. In general the line containing the point series, and the axis of the plane series, are different; but the condition is satisfied of itself when these lines coincide. A point series, (A, B, C), lies on a degenerate plane series, (α, α, γ), when the base line of (A, B, C) lies on the plane α; reciprocally, a degenerate point series (A, A, C) lies on a plane series (α, β, γ), when the axis of the plane series contains the point A. When both the point series and the plane series are degenerate, being (A, A, C) and (α, α, γ), the condition is merely that the point A shall lie on the plane α.

Two point series, whose base lines are not in one plane, determine what we call a *skew linear series*, consisting of sets of three lines of which no two intersect; such a set of three lines is obtained by joining the points, A, B, C, of any set of one of the point series, respectively to the points, A', B', C', of any set of the other point series. From any set of three lines, of which no two intersect, another set may be obtained, by taking any two transversals of these three given lines, meeting them, respectively, in A, B, C and in A', B', C'; then, determining upon the line ABC any point set, (P, Q, R), equivalent with (A, B, C), and upon the line $A'B'C'$ any point set, (P', Q', R'), equivalent with (A', B', C'); and, then, joining P and P', Q and Q', R and R'. The lines PP', QQ', RR' are then equivalent, as a line set, with the set of three given lines. It is clear that if, in this way, a line set (l', m', n') be equivalent with a line set (l, m, n), then the latter is equivalent with the former. It is necessary, however, if the word equivalence is to preserve its usual connotation, to shew, further, geometrically, that no logical inconsistency is involved in regarding two line sets, (l', m', n'), (l'', m'', n''), which are both equivalent, in the above sense, with a line set (l, m, n), as being equivalent with one another. In regard to this it may be remarked that the two common transversals of the six lines l, m, n, l', m', n', which enter into the definition of the equivalence of the line sets (l, m, n) and (l', m', n'), are not necessarily the same as the two common transversals of l, m, n, l'', m'', n'', which enter for the equivalence of (l, m, n) and (l'', m'', n''); it is therefore not part of the definition of the suggested extension of the meaning of the equivalence of two line sets, that the six lines involved should have two common transversals. It is necessary, for our purpose, to retain for the present the point of view of the descriptive Real Geometry; but it will be seen below, from a more general point of view, that there is, in fact, no logical difficulty. After what has been said in regard to degenerate point series, it is unnecessary to define degenerate skew linear series.

Reciprocally, a skew linear series may be defined from two plane series, whose axes have no point in common. A set of such a series consists of the intersections of the planes α, β, γ, of any set of one plane series, respectively, with the planes, α', β', γ', of any set of another plane series. If the lines forming a set so defined meet the axes of the two plane series, respectively, in A, B, C and A', B', C', the line set is equally determined by the two point sets A, B, C and A', B', C'. Take any three planes, λ, μ, ν, passing, respectively, through the lines of the line set, having a common axis, and, also, any other three planes, λ', μ', ν', with the same description; let ρ, σ, τ be a plane set equivalent with λ, μ, ν, and ρ', σ', τ' a plane set equivalent with λ', μ', ν'. If we regard the three lines, (ρ, ρ'), (σ, σ'), (τ, τ'), as equivalent with the original set, we shall only be giving, in a different form, the previous definition of equivalent line sets; for two equivalent sets of three planes, with the same axis, meet any base line in equivalent point sets.

A skew linear series is said to contain a point series when one of the equivalent line sets, of the former, contains one of the equivalent point sets of the latter. And, reciprocally, for the condition that a skew linear series should lie on a plane series. Thus, any two point series lying on a skew linear series serve to define the skew linear series (just as any two points of a line serve to define the line); and the skew linear series, which contains two point series lying on the same plane series, lies entirely on the plane series (as the line joining two points of a plane lies entirely on that plane).

It is clear that, if any line set of three lines, belonging to a particular skew linear series, be taken, and then any line, p, be drawn to meet any two transversals of the three lines taken, there is a set, of the skew linear series, of which the line p is one line. If l, m, n be three lines of a set of the series, and p a line, such that two transversals meet these four lines, respectively, in A, B, C, P and in A', B', C', P', then a point set, P, Q, R, can be found, on the former transversal, equivalent to A, B, C, and a point set, P', Q', R', on the latter, equivalent to A', B', C'. Denoting the lines QQ' and RR', respectively, by q and r, the three lines p, q, r are a set of the skew linear series. Hence it follows that, *of a given skew linear series, there is one set of which one line is an arbitrary line.* For, let O, O' be any two points of the arbitrary line; let ABC and $A'B'C'$ be two transversals of one set of the skew linear series; from O draw the line, OPP', meeting these two transversals, respectively, in P and P'. Denoting this line, OPP', by p, there is a set of the skew linear series consisting of p and two other lines, q and r. Now, from O, draw the line, OMN, meeting the lines q and r, respectively, in M and N. Then O, M, N is a set of a point series lying on the skew linear series. Another such set can, similarly, be found from the point O', say O', M', N'. The three lines, OO', MM', NN', are then a set of the skew linear series, of which the given line OO' is one line. For simplicity, the description has supposed all the lines involved to be in general positions relatively to one another.

The skew linear series, as has been indicated, generalises the properties of a line. There is however another aggregate of lines which also generalises the properties of a line. This we shall call a *line series*; it can coexist with, and must be distinguished from, a *skew linear series*. This last has been defined by means of two point series whose base lines do not lie in the same plane. If the base lines of two general point series, (A, B, C) and (A', B', C'), meet in a point, O, we may take, as sets of the two series, respectively, points, O, P, Q, on the former base line, and points, O, P', Q', on the latter base line. If then the lines PP' and QQ' meet in the point H, the three lines, HO, HP, HQ, are a set of three lines which meet in a point and lie in a plane. Taking any three such lines, l, m, n, we may define sets of lines equivalent with this set in a manner analogous to that followed above for point sets and plane sets; namely, we may regard the set l, m, n as equivalent with m, n, l or n, l, m, and as equivalent with any set, $l'\, m', n'$, of three lines, all passing through the same point and lying in the same plane, as l, m, n, when the three flat pencils (l', l, m, n), (m', m, n, l), (n', n, l, m) are related. The aggregate of all such equivalent line sets, of three lines in a plane passing through a point, constitutes what we call a *line series*. It is distinguished from a skew linear series by lying in a definite plane and passing through a definite point.

As before, we may then say that such a line series contains a certain point series, when the lines of one set of the line series contain, respectively, with proper regard to order, the points of one set of the point series. For this it is necessary that the base line of the point series lie in the plane of the line series. Reciprocally such a line series is contained in a certain plane series, when the lines of a set of the line series lie, respectively, in the planes of a set of the plane series. For this, the axis of the plane series must contain the point of the line series.

Such a line series may be determined by a general point series and a degenerate point series, of which one set consists of three coincident points. One set of the line series then consists of the lines joining this triple point to the points of a set of the general point series. Such a line series is also determined by a general plane series and a degenerate plane series, of which one set consists of three coincident planes. And, as has been indicated, the fundamental definition of a line series may be by two general plane series of which the axes lie in a plane.

Now consider three kinds of entities. The first kind consists of point series, and of points, the latter being regarded as degenerate point series. The second kind consists of skew linear series, and of line series, and of lines; the lines are here degenerate forms of either of the others. The third kind consists of plane series, and of planes, regarded as degenerate plane series. With appropriate definitions, we can then shew that, as the Propositions of Incidence, with Desargues' theorem, and Pappus' theorem, hold for points, lines and planes, so theorems corresponding thereto hold for the three kinds of entities. It is in the nature of the definitions of these entities that the Principle of Duality continues to hold also.

The detailed verification of the statement will only be sketched: the three kinds of entities will be referred as point entities, line entities and plane entities:

(*a*) Two point entities define a line entity, which is equally determined by any two point entities lying thereon. Two plane entities similarly determine a line entity. This is clear from what has been said.

(*b*) A point entity and a line entity determine a plane entity; a plane entity and a line entity determine a point entity.

It will be sufficient to consider the former statement. A point and a line determine a plane. A point and a line series determine a plane series, of which one plane set consists of the planes joining the point to the lines of a line set. A point, O, and a skew linear series, determine a plane series; for if u and v be two transversals of a set of three lines in the skew linear series, and p be the line drawn from O to meet u and v, the skew linear series, as we have remarked above, may be determined from a line set consisting of the line p and two other lines, q and r; and, if x be the line drawn from O to meet q and r, the planes xp, xq, xr are a set of a plane series containing both O and the skew linear series. It is interesting to verify that the line x so found is unique. The result involves that, through every point can be drawn a line to contain some point set lying on a given skew linear series (in another phraseology, through every point can be drawn a line to contain an imaginary point lying on a given imaginary line of the second kind; cf. von Staudt, *Geom. der Lage*, Erstes Heft, 1856, p. 81, no. 123).

Next, a point series, and a line, determine a plane series, whose axis is the line. A point series and a line series determine a plane when the base line of the point series is in the plane of the line series, namely this plane itself. When this is not so, let O be the point in which the plane of the line series is met by the base line of the point series; on this base line take the point set O, P, Q of the point series. If the point of the line series be at O, a plane series containing both entities is at once clear. When this is not so, let H be the point of the line series, and take a line set of the line series consisting of the line HO and two other lines, p, q, through H, lying in the plane of the line series. The planes, Pp and Qq, will meet in a line, say l, passing through H; these two planes together with the plane lO, are then a plane set containing both the point set O, P, Q and the line set HO, p, q; and they determine a plane series containing both the given point series and the given line series. Now consider a point series and a skew linear series. Let the point series be determined from a point set P, Q, R, and the skew linear series be determined from the two point series (A, B, C) and (A', B', C'). A plane series containing the last two series will contain the skew linear series, as we have remarked above; the problem, then, is, to determine a plane series containing three given point series (P, Q, R), (A, B, C), (A', B', C'). If (U, V, W) be any point series upon the skew linear series determined by (P, Q, R) and (A, B, C), then a plane series containing (U, V, W) and (P, Q, R) will contain (A, B, C); similarly, if (U', V', W') be any point series upon the skew linear series determined by (P, Q, R) and (A', B', C'), then a plane series containing (U', V', W') and (P, Q, R) will contain (A', B', C'). It will then be enough to find a plane series containing the point series (U, V, W), (U', V', W') and (P, Q, R). We can, however, so choose (U, V, W) and (U', V', W') that these lie on a line series, as we shall see; any plane series containing this line series will contain (U, V, W) and (U', V', W'). It will then be sufficient to find a plane series containing this line series and containing the point series (P, Q, R); which is the problem last considered. There remains then, only, to find the point series (U, V, W) and (U', V', W') in the appropriate way: (U, V, W) is to be a point series lying on the skew linear series determined by (P, Q, R) and (A, B, C), and (U', V', W') is to be a point series on the skew linear series determined by (P, Q, R) and (A', B', C'), and the point series (U, V, W), (U', V', W') are to be on a line series. We can suppose that the point P is not on the line AA'; for if P, Q, R be all, respectively, upon the lines AA', BB', CC', the point series (P, Q, R) lies on the skew linear series determined by (A, B, C) and (A', B', C'). The points, P, A, A', then, determine a plane; let QB, RC meet this plane in V and W, respectively, and let the line PA meet the line VW in U;

again, let QB', RC' meet this plane, respectively, in V' and W', and let the line PA' meet the line $V'W'$ in U'. The two point sets U, V, W and U', V', W', lying in a plane, determine two point series, and these lie on a line series in this plane, determined by a construction given above.

Lastly, consider two line entities. If these have a point entity in common, they both lie on a plane entity; reciprocally, if they both lie on a plane entity, they have a point entity in common. It will be sufficient to consider the former statement. A line and a line series having a point series in common, evidently, lie on a plane. A line and a skew linear series having a point series in common, lie in a plane series having the line as axis. Two line series of which the centre points are the same point, but the planes different, these meeting in a line p, contain respectively two line sets, p, q, r and p, q', r'; the planes (qq'), (rr'), together with the plane joining p to the line of intersection of the planes (qq') and (rr'), constitute a plane set containing a set of each of the given line series, and define a plane series containing these line series. Two line series, whose centre points are different as well as their planes, which have a point series in common, lie on a plane series whose axis is the line joining the two centre points. A line series and a skew linear series, which have a point series in common, lie on a plane series containing the line series and any other point series of the skew linear series. Finally, two skew linear series having a point series in common, lie on the plane series containing this point series and two other point series, one on each of the skew linear series.

(*d*) It follows from the preceding that three point entities, not on the same line entity, determine a plane entity, and that three general plane entities determine a point entity. Further, Desargues' theorem, as a consequence of the above Propositions of Incidence, is true of two triads of point entities. As regards Pappus' theorem, which now becomes a theorem that three point entities lie on a line entity, the conclusion is not so clear.

The preceding aggregates in the general Abstract Geometry. In addition to Pappus' theorem in its general form, there are many interesting geometrical questions raised by the preceding constructions. And, when these have been investigated, there is the question whether the entities serve the main purpose for which they are introduced; whether, for example, given four line entities, of which no two have a point entity in common, there exist two line entities having each a point entity common with each of the four given line entities. Such questions are not only of interest in themselves, but are of theoretical logical importance if it be desired to refrain from the extension of the Abstract Geometry indicated earlier in this chapter; the extension, namely, to be quite precise, in which a system of symbols is appropriate for which the equation $z^2 = c$ is soluble for all symbols c. But, if this extension be allowed, an account can be given of the preceding point series, line series and plane series, which not only sets at rest the logical questions, but, further, removes the apparent artificiality of the theory; such an account, by introducing clearness of conception, renders the geometrical discussion, of questions not considered above, much easier. To enter into this account, in detail, at this stage, would be to anticipate matters arising subsequently. Some indications, stated as the following examples, may, however, be desirable.

Ex. 1. For the cubic form

$$f = (x - ay)(x - by)(x - cy),$$

shew that the Hessian form, defined as

$$\frac{\partial^2 f}{\partial x^2} \frac{\partial^2 f}{\partial y^2} - \left(\frac{\partial^2 f}{\partial x \partial y} \right)^2,$$

is, save for a numerical factor,

$$(x - ay)^2 (b - c)^2 + (x - by)^2 (c - a)^2 + (x - cy)^2 (a - b)^2.$$

Ex. 2. Shew that if two triads of points on a line be given by

$$A = 0 + aU, \quad B = 0 + bU, \quad C = 0 + cU,$$
$$A' = 0 + a'U, \quad B' = 0 + b'U, \quad C' = 0 + c'U,$$

the necessary and sufficient condition that the ranges $(A', A, B, C), (B', B, C, A),$ (C', C, A, B) should be related, is that the two cubic forms

$$f = (x - ay)(x - by)(x - cy), \quad f' = (x - a'y)(x - b'y)(x - c'y),$$

should have Hessian forms whose (three) coefficients are proportional. The result can be obtained from what was proved above (p. 154), that the condition, for two ranges, $(P, Q, P + \rho Q, P + \sigma Q)$ and $(P, Q, P + \rho' Q, P + \sigma' Q)$, to be related, is that $\rho^{-1} \sigma = \rho'^{-1} \sigma'$.

Ex. 3. If ω be a symbol for which $(2\omega + 1)^2 = -3$, and the Hessian form of Ex. 1 be brought to the form $(x - ty)(x - uy)$, save for a factor independent of x and y, prove that, with proper distinction of t and u,

$$(t - a)(b - c) \omega^{-2} = (t - b)(c - a) \omega^{-1} = (t - c)(a - b),$$
$$(u - a)(b - c) \omega^{-1} = (u - b)(c - a) \omega^{-2} = (u - c)(a - b).$$

Ex. 4. In the preceding example, when ω is a particular one of the two symbols for which $(2\omega + 1)^2 = -3$, agreed upon, there is a distinction between t and u. And the point series of which a set consists of the points represented by $0 + aU, 0 + bU, 0 + cU$, taken in a certain order, can thus be associated with a particular one of the two points $0 + tU, 0 + uU$, which are determined when a, b, c are given; the other series, formed from the points of the same set differently arranged, will then be associated with the other of these two Hessian points. A similar remark is then applicable to a plane series, or to a line series. For a skew linear series an adequate statement can be made most conveniently only after the development of the theory of quadric surfaces.

Ex. 5. It has already been remarked, in the introductory article of this chapter, in anticipation of subsequent work, that, upon the plane curve called a conic section, a theory of related ranges can be set up; and the points of the conic can be put into unambiguous correspondence with the points of a line. We may then have point series upon such a conic, defined by equivalent sets each of three points, just as upon a line; and the two series, arising from the three points of a set taken in different orders, will be associated, respectively, with the two Hessian points of the cubic form belonging to any one of the equivalent sets. Although an anticipation, the geometrical form of this is so simple, that it may be mentioned: If the tangent lines to the conic at three points, A, B, C, be drawn, and meet the lines BC, CA and AB, respectively, in P, Q and R; then these points P, Q, R lie upon a line; and this line meets the conic in the two points, say U and V, which are the Hessian points of the three points A, B and C. Conversely, to an arbitrary line UV, there corresponds a series of sets A, B, C. With a further anticipation, it may be added, that when the line is what is generally called the line at infinity, and the conic is a 'circle,' the points A, B, C are the angular points of an arbitrary 'equilateral' triangle inscribed in the circle.

The points U, V are the common corresponding points of two ranges related by the fact that A, B, C, of one range, correspond, respectively, to B, C, A, of the other.

Ex. 6. If the theory of point series, etc., above given, is competent to represent all possibilities (as it is), then statements must be true which can be described, in a usual phraseology, not adopted here, as follows: An imaginary point lies on one real line; an imaginary plane contains one real line; there are imaginary lines of such kind that each contains a real point and lies in a real plane; and there are imaginary lines of another kind, upon which is no real point, through which passes no real plane. For the present we do not enter further into these questions; the precise discrimination of what is to be understood, in the Abstract Geometry, by a *real* point, and by an *imaginary* point, comes more naturally later.

Ex. 7. In the above account of point series, etc., regarded as aggregates of elements of the Real Geometry when this is extended by the adjunction of the postulated points, etc., we have remarked that three point series, of general position, determine a plane series. This is equivalent to saying that, if P, Q, R be three points of one line, and O, U, W be three points of another line, and O', U', W' be three points of a third line, all in general position, then three points, O_1, U_1, W_1, can be found upon the first line, equivalent to O, U, W in the sense explained, and three points, O', U', W', can be found upon the third line, equivalent to O_1', U_1', W_1' in the sense explained, such that the three planes, PO_1O_1', QU_1U_1', and RW_1W_1', meet in a line. We can regard the four points O, U, O', U' as fundamental for the threefold space of the figure; and, without loss of generality, we can choose the symbols for the points so that

$$W = O + U, \quad W' = O' + U', \quad R = P + Q;$$

if then the points P, Q be determined from O, U, O', U' by the syzygies

$$P = m_1 O + n_1 U + m_1' O' + n_1' U',$$
$$Q = m_2 O + n_2 U + m_2' O' + n_2' U',$$

it can be shewn that the line of intersection of the three planes, the axis of the plane series, contains the two points

$$pO + rU + p'O' + r'U',$$
$$-rO + qU - r'O' + q'U',$$

where

$$p = m_1 + m_2 - n_2, \quad q = -n_1 - n_2 + m_1, \quad r = m_2 + n_1,$$
$$p' = m_1' + n_2' - n_2', \quad q' = -n_1' - n_2' + m_1', \quad r' = m_2' + n_1'.$$

Ex. 8. It can be shewn that the general set of three points, O', U', W', equivalent to a set, O, U, W, upon the same line, in the sense explained above, the symbols of these being chosen so that $W = O + U$, $W' = O' + U'$, is found by taking

$$(O', U') = \xi (O, U),$$

where ξ denotes

$$\xi = \begin{pmatrix} a, & a-b \\ b-a, & b \end{pmatrix},$$

this notation meaning that

$$O' = aO + (a-b)\, U, \quad U' = (b-a)\, O + bU.$$

Here a, b are arbitrary symbols subject to $ab = ba$. Thus, if ξ' be a matrix of the same form constructed from a' and b', we easily find, from the law of multiplication explained above (p. 67), that $\xi \xi' = \xi' \xi$. We can write ξ in either of the two forms

$$\xi = \tfrac{1}{2}(a+b) + \tfrac{1}{2}(a-b)\,\epsilon, \quad \xi = b + (a-b)\,\tau,$$

where

$$\epsilon = \begin{pmatrix} 1, & 2 \\ -2, & -1 \end{pmatrix}, \quad \epsilon^2 = -3,$$

$$\tau = \begin{pmatrix} 1, & 1 \\ -1, & 0 \end{pmatrix}, \quad \tau^2 - \tau + 1 = 0.$$

Hence shew that the general point series contained on a skew linear series defined by the two (not equivalent) sets, O, U, $O+U$ and O', U', $O'+U'$, consists of sets of points, P, Q, $P+Q$, where P, Q are given by

$$(P, Q) = \xi (O, U) + \xi' (O', U'),$$

ξ and ξ' being any two matrices of the form above. Thus

$$P = aO + (a-b)\, U + a'O' + (a'-b')\, U',$$
$$Q = (b-a)\, O + bU + (b'-a')\, O' + b'U',$$
$$R = bO + aU + b'O' + a'U'.$$

This is a generalisation of the symbolic expression of any point on a line determined by two points.

BIBLIOGRAPHICAL

IT is possible to distinguish seven main sources of the theory of this volume: (a) The writings of the Greek geometers, and in particular the Collection of Pappus. Beside the particular result, here called Pappus' theorem, which is made to play an important part in the theory, the notion of the fourth harmonic point, the notion of six points in involution, developed by Desargues, and, essentially, what in the hands of Chasles and Steiner became the anharmonic ratio, here replaced by the theory of related ranges, are all to be found there, dealt with on a metrical basis; (b) The writings of Kepler and Desargues, who, both, from different points of view, regarded points at infinity as particular cases of finite points, and so made (or adopted from others) a momentous advance from the point of view of the Greeks; (c) The unpretentious but imperishable volumes of von Staudt. Though Desargues used the method of projection to prove the theorem we have here called by his name in the case of two triangles not in the same plane, he had no scruple in employing the metric methods of the Greeks for the case of two triangles in the same plane; not forgetting the inspiration of the great work of Poncelet, and his wonderful discovery of the imaginary elements of metric space, nor the far reaching character of the work of Chasles and Steiner, it is still to be remarked that for none of these did there exist any doubt of the obviousness of the notion of distance. It was von Staudt to whom the elimination of the ideas of distance and congruence was a conscious aim, if, also, the recognition of the importance of this might have been much delayed save for the work of Cayley and Klein upon the projective theory of distance. Generalised, and combined with the subsequent Dissertation of Riemann, v. Staudt's volumes must be held to be the foundation of what, on its geometrical side, the Theory of Relativity, in Physics, may yet become; (d) To von Staudt, however, the conceptions of the theory of aggregates, due to G. Cantor and others, were not clear; and his definition of related ranges, by means of harmonic nets, assumes that points of such a net are found everywhere on the line, as was pointed out by F. Klein. Thus it is necessary either to introduce into his scheme ideas which seem alien to his point of view, as indicated here in Chapter II, or frankly to recognise the need of an assumption. The discovery of the geometrical significance of this necessity is contained in an incidental

remark of H. Wiener; it is this remark which has led to the recognition of the importance to be attached to the assumption of
Pappus' theorem, or some equivalent assumption. It is a hazardous,
but it seems a natural forecast, that wide developments still await
the detailed investigation of the geometry in which Pappus' theorem
is not assumed, in which the appropriate algebraical symbols are
not commutative in multiplication. The definition of related ranges
given in this volume is not the same as that formally adopted by
von Staudt, but seems more in harmony with his general point of
view; for the possibility of the detailed development of this definition, the volume owes much to the work of F. Schur; (*e*) The
recognition that geometry does not deal with space in itself but
with definite figures, constructed from points assumed to be given,
in accordance with postulated laws of construction, and that these
constructions may be made in a limited accessible region. It would
seem right to associate these conceptions with the names of F. Klein,
Pasch and Peano; (*f*) To C. Segre and the writers of the Italian
school, is due the insistence upon the logical necessity and practical
utility of not limiting our outlook to space of three dimensions.
Though significant papers of Cayley, Riemann and Clifford are
to be named, the conviction of the increased geometrical insight
obtainable by the widened outlook is to be won only by experience
of the many striking successes in simplification and grasp which
have been obtained in this way. (*g*) Finally, the introduction by
Descartes of the equation to a curve, and the epoch-making work
of Poncelet (and of Laguerre) in regard to the imaginary elements,
though mainly of importance in subsequent volumes, are responsible,
ultimately, for the scope here suggested for the Abstract Geometry.
This extension is of quite vital importance for the development of
geometry; what is said thereon in Chapter III of this volume is to
be completed subsequently.

Some detailed references to these sources may be of interest:

(*a*) Exhaustive references to the Greek geometry must be sought
elsewhere. We may name M. Cantor, *Geschichte der Mathematik*;
the smaller *History of Mathematics* by F. Cajori; the German
Encycl. d. Mathem. Wiss.; Chasles, *Aperçu historique*, and Zeuthen,
Die Lehre von den Kegelschnitten in Alterthum (Kopenhagen, 1886).
The English reader will find the Prolegomena in Charles Taylor,
Ancient and Modern Geometry of Conics, Cambridge, 1881, especi-
ally instructive (pp. xvii—lv deal with the Greek geometry), and
will also consult T. L. Heath, *The thirteen books of Euclid's
Elements, translated from the text of Heiberg with Introduction and
Commentary*, 3 vols, Cambridge, 1908, and, by the same author,
Apollonius of Perga, Conic Sections, Cambridge, 1896. The actual
text of the theorem we have called Pappus' theorem may be seen

in Hultsch (Fridericus), *Pappi Alexandrini Collectionis quae super-sunt*, Volumen II, Berolini, 1877, and is as follows, the Latin version being by Hultsch:

pp. 884, 885 ἐὰν παράλληλοι ὦσιν αἱ ΑΒ ΓΔ, καὶ εἰς αὐτὰς ἐμπίπτωσιν εὐθεῖαί τινες αἱ ΑΔ ΑΖ ΒΓ ΒΖ, καὶ ἐπιζευχθῶσιν αἱ ΕΔ ΕΓ, γίνεται εὐθεῖα ἡ διὰ τῶν Η Μ Κ.

pp. 886, 887 ἀλλὰ δὴ μὴ ἔστωσαν αἱ ΑΒ ΓΔ παράλληλοι, ἀλλὰ συμπιπτέτωσαν κατὰ τὸ Ν ὅτι πάλιν εὐθεῖά ἐστιν ἡ διὰ τῶν Η Μ Κ.

si parallelae sint αβ γδ, et in eas incidant quaedam rectae αδ αζ βγ βζ, quarum αδ βγ concurrant in μ, et a quovis rectae αβ puncto inter α et β sumpto ducantur εγ εδ, quarum εγ cum αζ concurrat in η et εδ cum βζ in κ, rectam esse quae per η μ κ transit.

at ne sint parallelae αβ γδ, sed convergant in puncto ν; dicò tursus rectam esse quae per η μ κ transit.

(*b*) Kepler published a volume with the title *Ad Vitellionem Paralipomena, quibus Astronomiae pars Optica traditur* (Francofurti 1604, Cum privilegio S. C. Majestatis), of which the dedication to his patron, the emperor Rudolph II, is signed Joannes Keplerus, S. C. M[tis] Subjectissimus Mathematicus. In the four pages of this bearing the title *De coni sectionibus*, we find (pp. 93, 95) the two following passages, which indicate plainly enough how he was led to regard a point at infinity as a particular case of a finite point. The letters refer to a diagram given in his text.

Focus igitur in circulo unus est *A*, isque idem qui & centrum: in Ellipsi foci duo sunt *BC*, aequaliter à centro figurae remoti & plus in acutiore. In Parabole unus *D* est intra sectionem, alter vel extra vel intra sectionem in axe fingendus est infinito intervallo à priore remotus, adeò ut educta *HG* vel *IG* ex illo caeco foco in quodcunque punctum sectionis *G*, sit axi *DK* parallelos.

Oportet enim nobis servire voces Geometricas analogiae; pluri-mum namque amo analogias, fidelissimos meos magistros, omnium naturae arcanorum conscios: in Geometria praecipuè suspiciendos, dum infinitos casus interjectos inter sua extrema, mediumque, quantumvis absurdis locutionibus concludunt, totamque rei alicujus essentiam luculenter ponunt ob oculos.

Desargues' work with the title *Méthode universelle de mettre en perspective les objets donnés réellement ou en devis, ..., sans employer aucun point qui soit hors du champ de l'ouvrage* (à Paris 1636. *Oeuvres de Desargues, par M. Poudra*, Paris, 1864), reads (t. I, p. 82):

Quand les lignes suiet sont paralelles entr'elles et que la ligne de l'œil menée paralelle à icelles, n'est pas paralelle au tableau; les aparences des ces lignes suiet sont des lignes qui tendent toutes au poinct auquel cette ligne de l'œil rencontre le tableau....

His treatise on Conics with the title: *Brouillon proiect d'une atteinte aux éuénemens des rencontres d'un cone avec un plan* (Paris, 1639; *loc. cit.* t. I, p. 104) gives the definition:

Ordonnance des lignes droites.—Pour donner à entendre de plusieurs lignes droites, qu'elles sont toutes entr'elles ou bien 'paralleles, ou bien inclinées à mesme point, il est icy dit, que toutes ces droites sont d'une mesme ordonnance entr'elles; par ou l'on conceura des ces plusieurs droites, qu'en l'une aussi bien qu'en l'autre de ces deux espèces de position, elles tendent toutes à un mesme point.

(c) In the Vorwort to the *Geometrie der Lage*, by Dr Georg Karl Christian von Staudt, ord. Professor an der Univ. Erlangen (Nürnberg, 1847), we read:

Man hat in den neuern Zeiten wohl mit Recht die Geometrie der Lage von der Geometrie des Masses unterschieden, indessen gleichwohl Sätze, in welchen von keiner Grösse die Rede ist, gewöhnlich durch Betrachtung von Verhältnissen bewiesen. Ich habe in dieser Schrift versucht, die Geometrie der Lage zu einer selbstständigen Wissenschaft zu machen, welche des Messens nicht bedarf.

(d) H. Wiener, "Ueber Grundlagen und Aufbau der Geometrie" (*Jahresber. d. Deutsch. Math.-Verein.* I, 1892, p. 47), states:

Diese beiden Schliessungssätze (Der Satz von Desargues über perspective Dreiecke, Der auf das Geraden paar bezogene Pascal'sche Satz) aber genügen, um ohne weitere Stetigkeitsbetrachtungen oder unendliche Processe den Grundsatz der projectiven Geometrie zu beweisen, und damit die ganze lineare projective Geometrie der Ebene zu entwickeln.

v. Staudt's reference to the commutativity of multiplication of his Würfen is on p. 172, no. 268, of his *Beiträge zur Geom. der Lage* (Zweites Heft). A particular case of Pappus' theorem, considered with a view to the theory of similar triangles, is connected with commutativity of multiplication of the symbols by F. Schur, *Lehrbuch der Analytischen Geometrie* (Ers. Auf. 1898, Zw. Auf. 1912), Einleitung, p. 11. Cf. Hilbert, *Grundlagen der Geometrie*, 1899.

See also F. Schur, "Ueber den Fundamentalsatz der projectiven Geometrie," *Math. Ann.* LI (1899), pp. 401–409; "Ueber die Grundlagen der Geometrie," *Math. Ann.* LV (1902), pp. 265–292; and the volume by the same author, *Grundlagen der Geometrie*, Leipzig, 1909, pp. 1–192.

(e) For the many contributions of F. Klein to the foundations of geometry reference may be made to his forthcoming Collected Works. A general oversight may be found in the *Gutachten* pronounced for the first award of the Lobatschewsky prize, *Math. Ann.* L (1898), pp. 583-600.

Pasch's significant book has the title *Vorlesungen über neuere Geometrie* (Leipzig, 1882).

Peano's words may be quoted ("Sui fondamenti della geometria,"

Rivista di Mat. IV, 1894, p. 52): Ritenendo pertanto il concetto di spazio come fondamentale per la geometria, ne viene che non si potrebbe scrivere un trattato di questa scienza in una lingua che per avventura manchi di tali parole. Quindi non si potrebbe scrivere di Geometria nella lingua d'Euclide ed Archimede, ove appunto manca la parola correspondente al termine *spazio*, nel senso in cui lo si usa negli odierni trattati.

(*f*) The papers of Cayley and Clifford especially referred to are Cayley, "Chapters in the analytical geometry of *n* dimensions" (1843), *Coll. Papers*, I; Cayley, "A memoir on abstract geometry of *n* dimensions" (1869), *Coll. Papers*, VI; Clifford, "On the classification of loci" (1878), *Coll. Papers*, XXXIII; Riemann's paper, "Ueber die Hypothesen welche der Geometrie zu Grunde liegen" (see *Ges. Werke*), is of date 1854.

For references to C. Segre, see Bertini, *Introduzione alla Geom. proiett. degli iperspazi*, Pisa, 1907, pp. 1–426.

(*g*) Poncelet's statement of what constitutes in effect the Principle of Continuity in Geometry is given on p. 66, t. I, of his great work *Traité des Propriétés Projectives*, 2me Ed., Paris, 1865. For his discovery in regard to the imaginary (or ideal) points of circles, see pp. 48, 64, 380, etc., of that work, and, for the corresponding theorem in regard to spheres, p. 370. For Chasles' account of pairs of imaginary points reference may be made to Chap. V of his *Traité de Géométrie Supérieure*, 2me Ed., Paris, 1880. (See also the *Discours*, pp. 547–585 of the same volume, and his *Aperçu historique*.) Von Staudt's systematic justification of the use of imaginary elements occupies the Erstes Heft of his *Beiträge zur Geometrie der Lage*, 1856, pp. 1–129. In the Vorwort to this we read: Dass eine Ellipse oder Hyperbel durch ihre Brennpunkte und eine Tangente bestimmt ist, war schon den ältern Geometern bekannt. Das aber die Curve aus dem Grunde bestimmt ist, weil von ihr eigentlich fünf Tangenten gegeben sind, und dass also der erwähnte Satz nur ein besonderer Fall von einem allgemeinern Satze ist, ergab sich erst aus der Betrachtung der imaginären Elemente. A comparison of this with preceding formulations makes clear a notable advance in conception.

We pass now from these general remarks to some detailed references in regard to various points treated in the preceding pages. Explicit reference should however first be made to the two Cambridge Mathematical Tracts by Dr A. N. Whitehead, *The Axioms of Projective Geometry*, 1906, and *The Axioms of Descriptive Geometry*, 1907; also to Veblen and Young, *Projective Geometry*, Vol. I, 1910; and also to the *Bibliography of Non-Euclidian Geometry*, by Dr D. M. Y. Sommerville, St Andrews, 1911, which is a most valuable collection of exact references to geometrical literature extending far beyond its title.

pp. 6, 7. Desargues' theorem is proved as in the text by von Staudt. Desargues' proof for the case of two triangles in one plane is metrical.

p. 11. For the origin of the fourth harmonic point, see Ch. Taylor, *Geometry of Conics*, referred to above, p. liv.

p. 15. See the note to p. 60, below.

pp. 22–24. See F. Schur, *Math. Annal.* LI, 1899, pp. 406–409.

pp. 28, 29. This result, arising from Ex. 2 (p. 26), was so obtained by C. V. Hanumanta Rao, now Professor of Mathematics at Lahore.

p. 30. See K. Th. Vahlen, *Abstrakte Geometrie*, Leipzig, 1905, p. 129.

pp. 48, 49. See F. Schur, *Math. Annal.* LI, 1899, p. 405; F. Schur, *Grundlagen der Geometrie*, 1909, p. 37.

p. 60. The involution upon an arbitrary line by the six joins of four points, to which this work ultimately leads, is found in Pappus; see Ch. Taylor, as above, p. lii. The theorem was extended by Desargues to the intersections of a line by conics through four given points. *Œuvres de Desargues*, t. I (*Traité des coniques*, 1639), p. 268. For Desargues the relation is one among rectangles of segments on the line.

p. 61. Moebius, *Gesamm. Werke*, I, Leipzig, 1885, p. 443 (*Crelle*, III, 1828, p. 273).

p. 66. The symbols first given as example have of course the laws of combination of the complex numbers of ordinary analysis.

p. 71. In effect the symbols are here so introduced as to have the same laws of manipulation (save for commutative multiplication) as those of Moebius, in his barycentric calculus (1827. See *Ges. Werke*, I, p. 176), founded on metrical and statical considerations. In fact they were used as in the text, as the most appropriate representation of the geometry, before this was realised.

pp. 74 f. For such diagrams cf. F. Schur, *Math. Annal.* LV (1902), p. 281, or *Grundlagen der Geometrie* (1909), p. 51; Hilbert, *Grundlagen* (1899); Vahlen, *Abstrakte Geometrie* (1905); Veblen and Young, *Proj. Geom.* (1910).

p. 86. Moebius, *Ges. Werke*, I (Leipzig, 1885), p. 247, constructs the point $\alpha A + \beta B + \gamma C$, given $D = A + B + C$.

p. 96. Cf. Pasch, *Vorlesungen über neuere Geometrie* (1882); Peano, *I principii di geometria logicamente esposti* (Torino, Bocca), 1889; Peano, "Sui fondamenti della Geometria," *Rivista di Mat.* IV, 1894, pp. 51–90; Pieri, "I principii della Geometria di posizione," *Mem. Acc. Torino*, XLVIII (1898); also, Pieri, "Un sistema di postulati per la geometria proiettiva," *Rev. mathém. Torino*, VI (1896), and other writings of this author; Fano, "Sui postulati fondamentali della geometria proiettiva," *Giorn. di mat.* XXX (1892), pp. 106–132; Hilbert, *Grundlagen der Geometrie* (1899); F. Schur,

182   *Bibliographical*

Grundlagen der Geometrie (1909); Moore, "The projective axioms
of geometry," *Trans. Amer. Math. Soc.* III (1902); Veblen, "A system
of axioms for geometry," *Trans. Amer. Math. Soc.* v (1904); Veblen
and Young, "A set of assumptions for projective geometry," *Amer.
J. Math.* xxx (1908); Robb, *Mess. of Math.* XLII, 1912, p. 121; also
the writings of A. N. Whitehead, and of B. Russell; the German
Encycl. Math. Wiss.; and *Questioni riguardanti le matematiche
elementari, raccolte e coordinate da Federigo Enriques* (2 vols, Bo-
logna, 1912); and many other writings (see the Bibliography of
Sommerville, above referred to).

pp. 108–113. Cf. F. Schur, *Grundlagen der Geometrie*, pp. 16–18.

p. 125. In the hope that the geometrical reader may be willing
to examine the theory of an abstract order, given in the text, on
its own merits, references to, and the technical terms of, the theory
of the so-called arithmetical continuum, have been avoided. For
these the reader may consult authorities on the Theory of Functions.
The reference to Dedekind might quite well be accompanied by
references to Weierstrass, G. Cantor, and others.

p. 137. For Archimedes' axiom see the writings of Stolz; Hölder;
Enriques, "Prinzipien der Geom." *Encycl. der Math. Wiss.*; and
Veronese, *Fondamenti di geometria a più dimensioni ed a più spezie
di unità rettilinee esposti in forma elementare*, Padova, 1891, pp. 1–
628, which has given rise to an extensive literature.

p. 153. The argument of the text was suggested by F. Enriques,
Proj. Geom. Leipzig, 1903, Anhang, pp. 349–354.

pp. 165–175. Von Staudt's systematic treatment of imaginary
elements is in the first Heft of his *Beiträge zur Geometrie der Lage*
(1856); see, in particular, p. 76. An exposition of von Staudt's
theory was given by Ch. An. Scott, *Math. Gazette*, 1900, and
Bulletin of the American Math. Soc. VI, 1900, p. 163. For the
modified theory of the text the following may be consulted: Stolz,
Math. Annal. IV (1871), p. 416; August, "Untersuch. ü. d. Imag.
in Geom." (*Prog. d. Fr. Realschule in Berlin*, 1872); Klein, *Gött.
Nachr.*, 1872, p. 373; Lüroth, *Math. Annal.* VIII (1875), p. 145;
Sturm, *Math. Annal.* IX (1876), p. 333; Schroeter, *Math. Annal.* x
(1876), p. 420; Harnack, *Zeitschr. f. Math. u. Phy.* XXII (1877),
p. 38; Lüroth, *Math. Annal.* XI (1877), p. 84; Klein, *Math. Annal.*
XXII (1883), p. 242; Grünwald, *Zeitschr. f. Math. u. Phy.* XLV (1900),
p. 10; and, particularly, K. Th. Vahlen, *Abstrakte Geometrie*, Leipzig,
1905, pp. 163–169. The account given in the text is provisional,
and is to be completed in view of analytical developments given
later.

INDEX

to the pages of the text (not including the Bibliography).

PRINTED IN ENGLAND

AT THE CAMBRIDGE UNIVERSITY PRESS

BY J. B. PEACE, M.A.

Printed in the United States
By Bookmasters